QUEEN AIN'T GOT NO VEG!

A CATERER'S TALE

Alan Payne

With a foreword by Peter Alliss

VIGO HOUSE PUBLISHING

First Published in Great Britain in 2009 by
Vigo House Publishing
www.vigohouse.com

A CIP Catalogue of this book is available from
the British Library

ISBN 978-0-9563117-0-2

CHANDLER
BOOK DESIGN

Typeset in Garamond 11½pt
by Chandler Book Design,
www.chandlerbookdesign.co.uk

Printed in Great Britain by the
MPG Books Group, Bodmin and King's Lynn.

To Claire

Just see what trouble your degree could have got you into!

Bon appétit

This book is in memory of John Francis Coomb
and dedicated to the Morning Musketeers,
Roger de Pilkyngton, Bob Russell, Simon Bailey and
John Brooks as well as the hundreds of Payne & Gunter
People, young and old, permanent and casual,
regular and irregular who all succeeded over
many years in making me look very good.

CONTENTS

PETER ALLISS FOREWORD

The early days of my life in professional golf are a little hazy but for some reason one thing does stand out. What you may ask? Well, the life of the "outdoor caterer".

Back in the early 1950s catering at golf tournaments, and I suspect other outdoor events, was, to say the least, simplistic. Many golf professionals and those working at the event made "their own arrangements" as far as sustenance was concerned. Some of my fellow competitors brought sandwiches; others went into the clubhouse (if allowed) to have a bowl of soup and piece of yesterday's bread! All this changed with the coming of Augustus Payne.

It was in the late 50s, perhaps the early 60s, when I first spotted Gus (as he was more popularly known). He was on the shortish side, stocky, ruddy faced, with long sideboards, resplendent in a grand Norfolk style hunting jacket and trews made of some fine fabric, very shiny shoes and a cigar in hand. What on earth, I wondered, was he doing in our world of golf?

His business had been established in the Shepherds Bush area of London for some years. Gus was innovative and had given his personal attention to all sorts of catering opportunities. It was, however, his further involvement in the game of golf and all that entailed which pushed him to broader horizons.

As my game improved I moved up the golfing social ladder. I can't precisely remember when it was, but there came a day when I got an invitation to "go round the back of the caterer's tent about 11 o'clock. You'll discover something to your advantage!" I wondered what! Well, I was not renowned for hours spent on the practice ground, I wasn't due off the first tee until two o'clock, and so I had time to kill. With the help of a local Sherpa I found my way round to a tent of reasonably generous proportions and peeped inside. There was Gus entertaining some of the world's most famous sports writers, I thought I must be in the wrong place; there was Henry Longhurst, Pat Ward-Thomas, Peter Ryde, and one or two other younger looking faces which turned out to be those of Jack Woods of the News of the World and Desmond Hackett of the Daily Express. At this stage of my career I'd hardly spoken to any of these great men and now here I was, standing shoulder to shoulder, being offered a glass of champagne.

This became a regular feature at tournaments when Payne's were catering. It was a wonderful ritual, known to some as the "11 a.m. shift"! This meant a few invited people popped round for a glass of bubbly, a touch of gossip and a smile. I got to see behind the scenes and marvelled how his staff, who always seemed so jolly, could produce such tasty, interesting food, albeit by today's standards perhaps a little on the calorific side! I thought it was quite magical. Whenever I complimented Gus on the menu of the day, he replied, "Well, you know our motto, eat dirt cheap with Payne!"

I'm not sure today's Health and Safety boys would have given Gus's ensemble a total 'all clear' - the washing up and lavatorial arrangements were, to say the least, rather crude but, despite all that, I can't remember one single person going down with the dreaded "jim-jams", beriberi or the "caterers' two step". Outwardly Gus was a lovable figure with a host of friends, many of them I'm sure were "fair weather" but he didn't seem to mind. Who knows, something may come from that particular liaison and, after all, he only had a couple of large gin and tonics!

For some reason he changed his shoes at least four or five times a day - I wondered why, perhaps he'd had a touch of foot rot during his wartime service! Outwardly he gave the impression of being more London street trader and 'cheeky chappie' than Raymond Blanc or Fanny Craddock, but whatever his shortcomings, everybody was delighted to know Gus and whenever he appeared things somehow seemed to be better.

His son has put together this delightful book, which brings to life pictures of another place, another time. Alan had an interesting upbringing, very different from his father - although he didn't stay on and go to university, he went to a rather posh school, talked 'proper' and was fluent enough in a couple of languages to make him seem even more sophisticated and distant from the persona of his father. I wondered how on earth he would get on.

I felt the 'old man' was reluctant to hand over the reins, as most fathers are when it comes to their offspring following them into THEIR business. If you want confirmation of that, just ask Rocco Forte. Alan moved in and over the years created a much wider empire than his father could have ever envisaged, but with that came many pitfalls, but also a myriad of happy days and great and grand occasions, when he progressed from the golfing sporting scene to being responsible for the catering at a number of the world's greatest events.

Even if you know nothing about the world of catering, I'm sure you'll enjoy this book. For me it's brought back many happy memories of the Payne family and what they represented and presented in those far off days, not only in the world of professional golf but sport in general. Then, of course, there were the magical Air shows and grand banquets, a long journey from Shepherds Bush.

Time moves on but those of us lucky enough to have been there remember those times with great fondness.

Prologue

"How the hell did I get here?" I mumbled under my breath – at least I thought it was. For all my life I had been the invisible presence at other people's wonderful events and now for once I was centre stage.

I was in the Great Hall at London's Guildhall, scene each year of the Lord Mayor's banquet, and the State Trumpeters' fanfare was making the hair stand to attention on the back of my neck. Grand it may have been, but it was nothing new to me. As a caterer, I had heard the commanding tones of Bernard Sullivan, the toastmaster, many times over the years as he welcomed the assembled throng of the great and the good into the hall.

I had entered the vaulted gothic room with its banners of all the livery companies of the City billowing above, the tables covered in fine damask, bedecked with glittering silver and crystal, the six hundred guests on their feet, applauding the dignitaries to their places while the orchestra played Handel's "See the Conquering Hero Comes". The whole effect was, as always, quite literally breathtaking.

Then I stopped walking and heard the toastmaster pray silence for grace, and heard the grace being said and it was only then that I came to the full realisation that, unlike the last occasion I had heard this performance, when the applause had been for The Queen - this time it was for me. I was hosting the entire evening – presiding over the whole shooting match.

And then it came out.... but not quietly enough, for chuckles broke out and slowly rippled forth to the nearby tables. "Well you'd better write a book about it," someone said as the laughter died down.

1

Shepherds Bush

A bright but very chilly morning towards the end of April 1961 and I, a callow sixteen-year-old youth, was sitting on top of a five-gallon insulated tea-urn, almost lost in the surrounding pile of fifty other urns, as I proceeded to remove a year's worth of tannin from the interior, rims and lids, together with the mould that had grown during the previous winter's disuse. Annoying little bits of wire wool got stuck under my fingernails and my feet were losing any feeling.

I hadn't anticipated that both the urns and I would be spending the day sitting under the red brick gothic-arched entrance to The Old Fire Station in Pennard Road, Shepherds Bush on the cold stone cobbles that had been permanently stained during the previous century by the dung of Victorian horses as they were hitched up to the fire engines. Such was my first day working in the glamorous world of outdoor catering.

I knew the Fire Station well as it was the warehouse and cellars of my father's catering business. I had been brought here by him from a very young age on many a Sunday morning and played with my sister around the exciting rabbit warren of rooms while he got on with his work. Moreover, I had frequently worked here evenings, weekends and holidays during my time at school.

The many rooms down the stone steps in the cellar were now used for storing the hundreds of different drinks, dry stores and paraphernalia of a catering business. Since they had once been the dormitories of Victorian firemen, all the doors opened outwards with a push on the doorknob, speedier than turning, on the way to a fire. I was probably no more than five when I first saw these and they held a continuing fascination with my vivid imagination seeing dozens of moustachioed fire-fighters pushing these door handles while hurrying out to their possible death.

However this time my arrival was different. Now I was joining the firm full-time. To my father's employees, most of whom had either been in the army with him during the war, or worked alongside him pre-war at J Lyons & Co at Cadby Hall in Hammersmith, I was no longer the cheeky little boy who dived into the ice cream freezer. Now I was that dreaded pariah – the boss's son, claiming his inheritance.

In spite of being a less-than-brilliant student, I had thoroughly enjoyed my time at Latymer Upper School but I had left at Easter, having done two terms of my first year in the sixth-form. Many people have asked me, and certainly I've asked myself, why it was that a bright, inquisitive, competitive boy left school having just settled into his 'A' level courses. It is a question to which I have never been able to offer a satisfactory answer, least of all to myself, other than one word – Gus.

Gus – Gus Payne, my father, was an exceptional and memorable man. I realise that many people think their fathers exceptional but, certainly on the basis of the hundreds of letters my family received after his death, Gus entered and affected the lives of a tremendous variety and number of individuals.

One can trot out the usual clichés about exceptional men - "a character", "a one-off"; "larger than life" - and they all applied to Gus, yet still do him scant justice. The one thing that all such phrases miss is the overwhelming sense of enjoyment that he exuded. He was fun. He dramatically brightened the lives of almost everyone he met.

He was often described as Pickwickian and that certainly conveys a picture of him during the time I worked with him. Mutton-chop whiskers, almost foppish attention to his dress, always with a flower in his buttonhole, very stout and jolly; he could easily have featured in Dickens.

However, the best description of Gus was Churchillian. As well as looking like him, he enjoyed, and suffered from, many of Churchill's strengths and weaknesses. Certainly he had Churchill's charismatic leadership qualities, his energy, his ability to catnap and his facility for thinking on his feet and reacting to difficulties positively. He also had Churchill's knack of encouraging others to be equally positive.

Gus passed on some of these characteristics to me. I certainly catnap and have definitely proved that I can think on my feet. As far as leadership is concerned, I have without question the ability to get people to follow me, although my inability to take myself seriously means that I have always assumed that they did it because of sheer curiosity about where I was going, rather than charisma.

On the other hand, Gus also had Churchill's liking for the bottle – on almost the same heroic scale – and the same total inability to relate to his family. Sadly, like Churchill, Gus had received no affection at all during his childhood and therefore, also like Churchill, was quite incapable of giving it and, most especially, expressing it, to his children. I know that he loved me, because I have heard it from many friends since his death, but I never once heard it from him.

The main difference between Gus and Churchill was that while Churchill seemed to spend a lifetime creating adversaries of almost everyone, Gus made friends. That, more than anything else, was his life's work, making friends. Everyone responded to his contagious friendliness; he gave of himself totally from the instant he struck up an acquaintance, in a restaurant, or at one of the events where he was catering. Gus's most notable attribute was the fact that everyone who met him felt that they were the only person who existed for him at that moment.

He always appeared to have time for his friends and acquaintances; he would do anything, give anything and was therefore an easy touch for the inevitable freeloaders. There was, however, never time for his family – only once did he come to my school - and that was to organise a catering function.

He never tried to be anything he wasn't. He was never ashamed of being a West London lad. Quite simply he was Gus Payne from Shepherds Bush; acceptance or not was the other person's problem. He was equally at home with Old Etonians or the fellow urchins he grew up with in the Bush Market. Gus believed that you should, "treat a duchess like a charlady and a charlady like a duchess".

His conversation was usually colourfully punctuated with explosive guffaws of laughter that were totally infectious and normally resulted in the complete uproar of everyone around. This laughter almost invariably culminated in a fit of coughing brought about by a semi-permanent Ramon Alones Petit Corona. On the rare occasion a cigar was not present, a Sobranie Virginia in the end of a silver-and-black cigarette holder took its place.

He was not flash; he was elegant. With Saville Row suits, always with a fresh carnation in the buttonhole, Jermyn Street shirts and shoes hand-made in St James's, he would never have been taken for a lad from the 'Bush'. He was a boulevardier, out of his era. His natural habitat was the West End of London's restaurants and dining clubs.

We, his family, while not seeing a lot of him, were just as enraptured by him as any of his friends. I totally hero-worshipped him. I loved him dearly, to the extent that, many years after his death I often found myself on my own, very early in the morning, in my office, in floods of tears having just caught sight of his picture and realising just how much I missed him.

He was a lovely, lovely man, enormously engaging, a pleasure to be with, and adored by everyone, especially me, but - he was a bully. I don't think he meant to be, but he was. The idea that I was taken to all the catering events during my school holidays to work was not considered out of the way, by me or anyone else. Neither was the fact that I used to go to work at the various cocktail parties of his friends or waited tables and washed up at dinner parties, probably from the age of eleven or twelve.

For the period of my time at Latymer it was an exceptional weekend during the summer when I would not be expected to work at some event or other and, as a consequence, I never played any sport enough to become any good at it. As well as being bad for sport, it was not a regime that encouraged academic activity, particularly since Gus had always been less than diligent at following my school progress.

When, largely due to the intervention of one of the

masters at Latymer, I started to do my 'A' levels in the sixth form, I loved it. So why leave after the first two terms? Quite simply, it was impossible to swim against the tide of Gus's personality. He wanted me to join the company, the summer season was about to start and I found myself at work in Pennard Road. It was as simple as that.

Years later my mother told me that she had tried to persuade Gus to allow me to continue on to university but he was afraid that, had I left home and gone to college, he would have lost me and I would never have joined his business – "How you gonna keep 'em down on the farm now that they've seen Paree?"

Who knows whether he was right? I think that probably I would have come back, and brought some 'learning' with me. I regret more than anything not having had the time to myself as subsequently it took many years, a great deal of conscious effort and, most especially, my wife Judy, to extract me from the total domination of Gus's influence.

However, when I arrived that April morning at the Pennard Road 'Fire Station' the initial resentment was palpable, although I certainly wouldn't have known at the time what that meant. I only knew that a number of the staff, who previously had joshed around with me happily when I was working at tennis tournaments during the Easter holidays or helping in the kitchen at golf tournaments during the summer, were suddenly acting very differently.

What I didn't realise when I first started work was that, for the junior ones, I was an upstart who was probably going to jump over them, and, to the senior ones, my father's fellow directors, I was a distinct threat, albeit a few years off.

Most notable among the latter was my father's second in command, Alan Cox, a powerful, six-foot-two ex-sailor, who had been on the arctic convoys to Archangel during the war. He had assumed that he would take over from my father and he did his best to make my life very unpleasant, obviously to see if he could make me give up.

Had he succeeded, I wouldn't, of course, be writing this, but this was the start of a very tempestuous relationship over the next twenty years. Besides anything else, because of our relative sizes and ages, he was always 'Big Alan' and I, henceforth, to all and sundry, was 'Little Alan'.

To make matters worse, my father, in order to prove no favouritism, made it very clear that I was to be given no quarter, to be awarded the most unattractive jobs, and to justify twice over any promotion I received. Naturally some of the other staff took their cue from him and so my early days at work were pretty wretched. Hence the pile of urns saved up for my arrival.

Looking back I was astonishingly stoical, particularly as I didn't know what that meant either. I just got on with it, did whatever was required, joined in the endless leg pulling (for which I was inevitably the butt) that a close coterie of fellow workers travelling in each other's company invariably generates, and thus slowly won acceptance and eventually respect.

So, on that sharp and lovely spring day I travelled by trolleybus from Acton and arrived at Pennard Road, which was to be my base for the next three years. This was one of two premises in Shepherds Bush belonging to Paynes Restaurants; the other was across the main Uxbridge Road, opposite the north end of the market. 204 Uxbridge Road had once been one of the eponymous restaurants but by this time had been turned into a delicatessen, with our kitchens in the basement and two floors of offices above.

The company's name was a misnomer as, by then, we no longer had any restaurants. Our activities at that time were primarily cocktail parties, fashion shows and weddings and, during the summer, outdoor events such as agricultural shows and golf and tennis tournaments, or even, occasionally, what my father referred to as the lowest form of human life - Conservative Party Fetes.

We also had a flourishing hire business supplying many companies with the equipment necessary for their own one-off functions and also providing a service to other caterers.

The most notable of these during the sixties was Prue Leith, who used to roll up to Pennard Road in her Renault 4 to collect the equipment and linen for dinner parties for her fledgling catering company.

As a shy young warehouse lad, I remember thinking, while I loaded the equipment into the car of this very attractive and slightly older woman, that she must have been a very clever young thing to have started and be running her own business. She certainly proved me right, although forty-plus years on she now appears, quite patently, much younger than me.

My first evening function as a full-time employee was a members' evening at the Royal Geographical Society in Kensington Gore. We undertook these functions almost every week on Monday evenings during the winter and spring and, because of the noise we inevitably made setting up, everything had to be ready before the meeting and lecture started. We then spent two or three hours sitting in the dark, dingy hole-in-the-wall storeroom that we had turned into a still-room with temporary Calor gas boilers to make the tea and coffee for the three or four hundred members of the Society. Once we had served their interval refreshment we had to wait until the second half of the lecture had finished before we could leave with our equipment.

This was an early sign that a great deal of my life as

a caterer was going to be spent 'hanging about', and, if I wasn't to be bored senseless, I was going to have to find a way of occupying myself. There was no problem with this at the Royal Geographical Society as this first evening I discovered a wonderland of artefacts, maps and globes, reeking of two hundred years' worth of furniture polish and steeped in the history of the great Georgian and Victorian travellers who brought their tales of derring-do back to their fellow members. There was more than enough to keep me intrigued every Monday evening, wandering round the whole building while the members were in the lecture theatre. Pulling the books and maps from the library shelves I was transported to being a part of the heroic 'Boys Own' exploits of Captain Cook, Scott and Livingstone. The time I spent in the company of the explorers was all the more beneficial when I realised the alternative was to sit with the porters and drivers, watching the Calor gas boilers bubbling away, and listening to the discussions about the relative merits of Queens Park Rangers, Chelsea and Fulham Football Clubs.

The RGS was merely the first of many wonderful buildings I found my way into while waiting for things to happen at catering functions, and over the years I was going to find treasure trove aplenty in some entrancing places that often made the 'hanging about' the best part of the job.

The reason I was spending my first week cleaning out the 5-gallon urns back at Pennard Road, was that they were all needed by the end of the week when we were off to Bournemouth for the British Hard Court Tennis Championships. Our away-season was about to start, fitfully to begin with but, by the end of May, we would be on the road for three months, pretty well non-stop. So everything had to be ready for the off.

The Friday of my first week eventually arrived and I received my first wages; two pounds ten shillings. I can vividly remember what I did when I got the manila Seal-Easi Wages Packet. Taking one of the several thousand Apostle-headed teaspoons in the warehouse, I went straight out and bought a tin of Nestlé's condensed milk, which I loved. I then walked over and sat on Shepherds Bush Green and I ate the whole tin with the teaspoon - the first blossoming of my life as a gourmet. I was very ill, and have never been able to face the stuff since.

Bournemouth Tennis was one event I knew well as I had been there during the Easter holidays from school since I was ten. It was here that I first came across our temporary staff. As well as some locals we employed from Bournemouth, we also took a number with us from Shepherds Bush, regular casuals who worked all year on and off and some who just came to this one event, taking a week off from their normal jobs. They were a motley crew but great fun. As a schoolboy I found it magical.

We would all foregather in front of the warehouse very early on a quiet and eerie Sunday morning in Pennard Road. Slowly the entire charabanc of three large vans with caravans behind, a mini-bus and five or six cars was made ready. There was always an unseemly scramble with much of the squealing and shrieking, "I don't want to sit next to 'er" and "No you sit in the back, I'm sittin' ere". I was not aware at the time but this was going to form the musical backdrop to a large part of the next decade. All the occupants were eventually settled in the right vehicles and we started off in convoy down the old A30 towards Bournemouth.

I always tried to get in the biggest van, because it was driven by little Tommy Potter, who was five foot two and shrinking, with a shrivelled walnut of a face, and a

Woodbine that had been in his mouth since the Old Queen died, and he was very funny. I also liked the big Bedford because it was exciting.

This van, which I was to drive myself in years to come, was a serious menace. It had originally been an army wagon during the war, sixteen years earlier. After the war a custard powder company had converted it to carry a very large amount of custard powder, which, of course is extremely light. The custard powder company went broke and my father bought the van, only to fill the enormous space with trestle tables, Calor gas bottles, boxes of china, cutlery and silver, almost none of which could be called light.

The result was that this aged vehicle with tractor-style tyres was always dangerously overloaded and caused enormous queues behind it on the pre-motorway roads as it laboured up hills, towing a caravan on the back, and then pelted recklessly down the other side.

The cabin was tiny, made totally from wood with the roof just a couple of inches above our heads. This was just possible for Tommy but for the others it was excruciating. Even for Tommy the process of changing the ancient gears, of course requiring double-declutching, usually resulted in bashing one's elbow on the back of the cabin and, in reaction to the considerable pain, shoving the gear stick forward, usually into the wrong gear. Given the overloading and the number of hills on the A30, the gears were being changed every few minutes. As a kid I knew none of this except for Tommy's swearing at the gears, which I thoroughly enjoyed.

A high spot for me was when the entire column of colourful vehicles, with their even more colourful occupants, stopped at a transport café in the wide High Street of Hartley Wintney for a full fried breakfast and a pint mug of very strong tea. In the days before the motorways, these cafes, usually referred to as 'greasy spoons', were the traditional meeting places of the trunk roads and the banter and backchat was always a revelation to the young schoolboy that I was at that time.

After breakfast, we were on our way again, stopping also for lunch before we arrived at the tennis club. What nowadays takes an hour and a half was then nearly a full day's drive. As a kid I had found the whole week wonderful, selling ice cream from a Wall's Ice Cream bike, getting up to mischief with my dad's staff in the evenings; it was a great holiday, but now in 1961, this was for real. What was it going to be like as a fully-fledged worker?

A few days before I was due to find out, I was at home and found my father in the grip of helpless laughter. He couldn't speak but handed me a letter and parcel he had just received. They were both from Bethan Jones who I knew from my times at Bournemouth. She was one of our regular local waitresses who worked with us for only one week of the year at the tennis at Bournemouth, and had done for a dozen or more years. She was a very devout woman and would respond to our hellos each year with, "Hallelujah!" and, when asked what sort of year she had had, would cryptically inform us, "Do you know, I've worn out four tambourines this year?" With this background, the first part of letter should not have come as a surprise. It read:

Dear Mr Payne

I will not be working with you at the tennis at the West Hants Club this year, as I have decided to devote the rest of my life to God and to do His Works. You have been very good to me over the years and I fear I have not repaid the goodness. I therefore have sent in a separate package, all the cutlery and things I took from you over the past ten years. My apologies and God Bless you.

Yours in the name of the Lord

Bethan Jones

The accompanying parcel contained seventy covers of silver cutlery, three hundred and fifty pieces, together with silver cruets, mustard pots, milk and cream jugs and coffee pots; a veritable pirate's hoard. Truly, the Lord's work was wondrous to behold.

My first away-week at the tennis as a full-time member of staff turned out to be as much fun as it had been before. My relationship with all the temporary staff had not been affected at all by my new status, particularly when they realised that I was prepared to work as hard as, if not harder than, anyone else.

The main difference I found from my new position was that as a schoolboy on holiday I had slept in my father's caravan, but now as 'one of the lads' I was expected to sleep with the other youngsters, porters and trainee chefs on camp beds under the concrete stadium. Pretty grim and cold (one year it snowed) especially since I was expected to wake up first and light all the boilers in the makeshift kitchen that shared the space with us.

One thing I learnt from this first week was that, although some of the staff gave me a hard time, there were a number of waitresses and bar staff who were absolutely the opposite, overly sweet and fawning, saying embarrassingly nice things about me to my father. I found this excruciating and so, I'm sure, did the old man. The bullies, including my father, were much easier to deal with.

At the tennis tournament I was teamed up with Doris who was very much one these 'crawlers'. We were running a very, very busy bar – a real baptism of beer. Each time a set or a match finished we were deluged with a mob of jostling, argumentative spectators in a desperate hurry to get a drink and to return to the next match. We coped very well but at the end of the day Doris, to my cringing embarrassment, said to my father, "Mr Payne, little Alan is such a brave young man - he doesn't know the meaning of the word fear".

My old man guffawed and told her not to be such a bloody fool. But she was quite right of course, along with stoical and palpable, fear was one of many words of which I didn't know the meaning.

One of the bright spots that first week away in Bournemouth was the day when one of Gus's friends from the golf tournaments came as his guest for lunch. This was my first meeting with Peter Alliss, a star Ryder Cup golfer and already showing the gift for talking, amusingly and endlessly, that was to make him such a success as a television commentator.

Since in age he was almost exactly half way between my father and me, it was one of the most notable instances of one of the old man's friends becoming my friend. This was helped in no small measure by the fact that he also had suffered from the burden of following a famous father, Percy Alliss, a professional golfer of considerable fame at a time when there was little celebrity in the sport. Some twenty years later we were joined by three other similarly afflicted friends to form a short-lived but vastly enjoyable Society, which we called S.O.D.S. - Sons of Distinguished Sires.

The other three were Gerry Kunz who was running the Motor Show at that time and whose father was the world famous pianist Charlie Kunz; Stanley Holloway's son, John, a customer of mine at Smiths Industries; and Bill Cotton who was Managing Director of BBC Television and the son of Billy Cotton, he of the famous Band Show.

For a year or so we had occasional meetings at unusual places that one or other of us could organise, such as the lunch we had watching Peter record a couple of his "Round with Alliss" programmes. I seem to remember extremely convivial meetings over rather a lot of food and drink, endlessly retelling tales of the dreadful deeds of our fathers. We also set about the business, quite earnestly, of developing a set of rules that would exclude Mark Thatcher.

I am very sorry now that I didn't persevere with S.O.D.S because during the next couple of decades I have met a great number of people who would have qualified by dint of their outrageous fathers. However, political correctness would probably now insist on it being called OOFF – Offspring Of Famous Folk.

All this excitement was going to happen to me years hence and I, of course, knew nothing of it when, as a young sprog, I first met Peter at Bournemouth.

After enjoying my week at the tennis, coming back to the Pennard Road warehouse was pretty miserable. Luckily I did not spend a very large part of the summer there as within a few days we were back on the two circuits that formed the bulk of the company's business, Golf Tournaments and Agricultural Shows. At the time I joined the company, we looked after about a dozen golf events each summer, which was to rise to a peak of fifteen or sixteen by the mid- seventies.

Agricultural Shows on the other hand were just about at their maximum of twenty-five or so, a lot of them one-day events, which we would progressively pull out of as more profitable business took their place. The biggest was the Royal Show which would require the presence of the whole company, but the regional three-day events, which ran on one after another, in May and early June, were looked after by Alan Cox with two or three waitresses, a driver and a chef and, from 1961, me.

It was a merry band, which joined up with all the other contractors for mutual help and some considerable socialising in the evenings at each event, rather like a circus. We would leave one site in the early evening and in a mid-sized van (not the custard-powder Bedford) with a caravan behind would travel to the next showground, park the caravan and go into town to meet our fellow show-people for a drink or a meal.

So I was sitting in the front passenger's seat of the van with Alan Cox driving as we made our way from the Three Counties Show at Malvern to the Royal Welsh Show at Builth Wells in deepest Wales. In order to be prettied up for meeting the others in the pre-arranged pub in Builth, the girls were readying themselves in the caravan rather than sitting in the back seats of the van.

The van was struggling to cope with all the equipment on board as well as the caravan behind as the route was very hilly. We just made our way to the top of one particularly steep hill and Alan was fighting to limit the speed on the downward turn when he saw that someone was overtaking him, in spite of it being a very narrow road. He was going to use some of his well-honed nautical terms on the driver, when he realised that it was not another vehicle, but our own caravan – with the three girls on board, which had sheared off from its coupling to the van.

Helplessly, we watched as the caravan careered from side to side across the road before turning on its side in a ditch, a complete write-off but fortunately the occupants, while very far from being prettied-up, had only suffered mild cuts and bruises as well as some shock, which took quite a few gin and tonics from the back of the van to remedy.

These were, of course, days long before mobiles or any other means of communication and we just had to wait until the next vehicle arrived on the scene to start the process of getting the RAC to get us to Builth and find somewhere to sleep, since we no longer had a caravan. As we drove on down the hill we were terrified to discover that a mere hundred yards further on, the side of the road was a fifty-foot drop.

The show coterie sorted out accommodation for the night and the next day I went with Alan to buy a new caravan, which the girls slept in the following night – Alan and I slept in the back of the van - only to wake up the next morning to a totally flooded show-ground with water up to the steps of the new caravan. I was starting to realise that life as a peripatetic caterer was going to be anything but dull.

2

GUS AND LILA

At the time I joined the company, Gus was really in his pomp. Fifty years of age, the muscles from his very sporting youth had, to put it kindly, softened and expanded, helped no doubt by his fine taste in food, prodigious appetite for alcohol and a very gregarious but sedentary lifestyle whilst building the business. He thoroughly enjoyed the product he served.

He cut a very successful figure; to be more precise, his tailor, Mr Pederman, cut a very successful figure for him, as he always managed to appear portly, not fat. He was the absolute embodiment of the good life and the bonhomie that he was selling. He had established his company as a leading one in event catering with a unique stranglehold on the world of professional golf, which was growing at a considerable rate as a valuable way for sponsors to befriend their customers. This was about to explode with the arrival of the Big Three from the American Golf Tour, Palmer, Player and Nicklaus, together, of course, with their groundbreaking manager, Mark McCormack, who Gus worked with from his very first days in the UK.

By the time I started work at Pennard Road, Gus had established a routine for the whole company that fitted his personality and requirements. Since I used to begin each day between seven-thirty and eight, along with the rest of the team, I was well into my day's activity when Gus would draw up in his car, and little Tommy Potter would go out to meet him. A brief discussion about what was required later, accompanied by the inevitable guffaw and cough at one of Tommy's jokes, and Tommy would take the car to be filled up and washed.

I would carry on with whatever I was doing, counting teaspoons, packing china, loading or unloading vans, while Gus came into the yard, bringing with him the sumptuous aroma of the first of the day's Petit Coronas. After a short conversation with, the warehouse manager, Jimmy Wilkins, and perhaps a nod to me, he would be off over the road to 204 for his morning's office work.

This always comprised the minimum amount of paperwork, a few letters dictated to Betty, his secretary, a tall, handsome lady, with big fashionable 'secretary-in-Doris Day-film'-type glasses, a wicked sense of humour and who was more than a match for Gus. Like everyone, she adored him but had no rose-tinted glasses about his shortcomings; it was her job to cope with them, and to make sure the world outside only saw the Gus that they, and he, wanted them to.

There then followed a couple of hours of pell-mell telephone calls with Betty alongside him preparing call after call on the two lines. It was a pretty impressive operation to observe, as over the previous few years she had developed an intuitive feel of where he was on the current call and when to start the next. Neither Gus, nor the recipient was ever kept waiting.

No later than twelve o'clock, Gus would arrive back over at Pennard Road, and either he would drive himself or, more usually, Tommy would drive him on the next, most critical, part of his day. The West End awaited him, as I discovered a year or so later when I assumed the role of chauffeur on many occasions. First stop, was to pull up outside the Dorchester Hotel for Gus to go to the flower shop and select a new, clove carnation for his buttonhole, specially cut by Mr Pederman in a crescent so that it could take the bulb of the flower without distorting the lapel, together with a couple of spare blooms for later in the day.

Next call was to one of four clubs where he was a member, for lunch with someone who was of interest, possibly not immediate interest, but maybe sometime in the future. Each lunch was an investment. Gus never considered that it could possibly be a waste of time, even if the prospect he was entertaining didn't develop - so what. It had been an enjoyable lunch and he had learnt something that could well bear fruit at a later date. Whatever he learnt was invariably scribbled down on a scrap of paper that was thrust in the side pocket of his suit to be unfailingly retrieved the next morning, no matter how late or lively the previous night. He was never hard-nosed about whom he entertained and sometimes was taken advantage of, but much better to give away a few meals, which he enjoyed anyway, than miss one opportunity.

The four clubs, along with many others, knew him well, as lunching and drinking at them had been his real marketplace, even when he had first started the fledgling business immediately after the war. This was where he demonstrated what the word hospitality meant. The clubs, none of which still exist in their previous state were, in order of use, The Belfry Club (now Mosimanns), the Hanstown Club, the Wellington Club and, most appropriately, the Eccentric Club.

If he was driving himself – this was pre-breathalyser – the doorman would take his car and return it when required. Gus would go into the bar and without any signal a Booth's High & Dry gin with Schweppes Tonic would be placed before him. He had made the dramatic change from Gordon's Gin in 1959 when he had undertaken the catering for the opening party of Booth's new distillery in Clerkenwell, at which I had worked after school at fourteen years of age. Needless to say, he still remained great friends with the Gordon's directors and would still consume their product without too much demur. His guest would be brought over to him by the Maitre D' - he always arrived before them - and they would, in current parlance, be well and truly schmoozed for the next two to three hours.

That this all worked brilliantly is shown by the chain of events that landed him the biggest contract of his career, which lasted for twenty-five years, for the Dunlop Masters Golf Tournament.

Gus's mother, who originally ran the business, had banked with the local branch of Barclays and Gus continued this. However, since his requirements were on a grander scale he made contact with one of the senior managers, Calvin Dubois, who was subsequently to become Barclay's main man in London.

A lunch revealed that Calvin was president of Sutton Lawn Tennis Club; Gus took over the catering for the club's annual tournament and social dinner dances. He also subsequently undertook the weddings for the daughters of several members. At one of these he met a guest who lived next door to the host; the neighbour was a man called

Frank Smith, who turned out to be the boss of the Dunlop Sports Company.

A lunch with him elicited the information that they had a very large golf tournament, (which Gus, of course, already knew), and that Gus had better meet Eric Hays, the man who ran Dunlop Golf.

The lunch that followed with this delightful, genteel, eccentric Scot, who lived with his mother high up on the Sussex Downs, drank easily as much as Gus, and was one of the most influential men in golf, ended in Eric being decanted from Gus's car into the train at Victoria, a life-long friendship and business relationship having been established, with Paynes Caterers having its cash-flow assured for the next two decades.

Eric's words at the time to his boss, Frank Smith, were, "If he looks after our guests the way he looked after me we have nothing to fear". Having similarly revelled in Gus's hospitality, Frank was in no mood to argue - and Eric was quite right.

So, after the all–important lunch, Gus's daily progress would continue to one of the West End's many drinking clubs, which existed at that time to circumvent the arcane licensing laws that had been introduced to keep the armament workers sober during the First World War and were not to be amended until the nineteen-eighties.

A couple of hours here in the company of an agreeable series of similarly eccentric characters and then Gus would descend upon one of his many fine friends who were the suppliers of his daily intake, the directors of the grand old companies which made up The Distillers Company, or, as it was universally known in the trade, DCL.

The companies, a dozen whisky, two gin and one vodka, that came together in the nineteen-twenties as DCL, still operated as separate trading entities, each with its own very grand West-End office, a well fed and watered board of directors and separate sales force, all with extravagant expense accounts. They were destined to be the subject of the profitable pillaging of Guinness, twenty-five years later, under the direction of the infamous deadly Ernest Saunders.

In the early sixties their boardrooms were a series of well-stocked boltholes for Gus to round off virtually every day. The directors of these companies welcomed him as a signal that it was time to sign-off all their correspondence and start sampling their product with him.

An hour or two of this and it was time for Gus to return home to Acton. At least a couple of times a week he would then get changed into his exceedingly elegant dinner suit, bedecked with gold Albert chain across his very round front, to go to a trade dinner or, on Monday evenings during the winter, entertain a couple of guests with an evening's dinner and boxing at the National Sporting Club, run at that time by Charles Forte at the Café Royal. Gus was certainly making the most of his life during that period and enjoying it all immensely - and why not? The first fifty years had certainly been far from easy.

His mother had been a very formidable lady from Abergavenny, called Alice Holl, who had come up to London towards the end of the nineteenth century and worked in catering in the West End and eventually, with a small inheritance, acquired her own coffee stall on Wandsworth Common. It was, in the vernacular of the time, 'a pull–up for carmen'.

Somehow she met a ne'er-do-well from Shepherds Bush called Wilfred Payne, whom she married and kept for the rest of his life. Alice moved into a flat with him and his mother in Wood Lane and trundled her stall from

Wandsworth to Shepherds Bush Market.

I knew Wilfred very little – he died when I was six - and he was not talked about a great deal. He was quite patently not held in very high regard by anyone in the family – the black sheep from Shepherds Bush, perhaps - although he most certainly must have had something about him. From all that I heard, he never, ever worked, was up in the West End every evening until the wee small hours and was a gambler and womaniser of some distinction. Considering the fact that he looked a dead ringer for Old Man Steptoe, (who, of course, came from the other end of Shepherds Bush Green), that was quite an achievement.

The only thing of his that was passed on to me at the time of Gus's death was a black silk opera top hat, the sort that pops-out and folds flat, which apparently Wilfred always wore 'up-West' – not bad for the indolent husband of a coffee stall holder in Shepherds Bush Market. Why Alice put up with him I have no idea, as she was a very strong-minded woman, very hard working with a dominating personality. She was quite obviously the main source of income, as any winnings Wilfred might have made he would have been put to nefarious use, so she could have chucked him out at any time. Certainly their life together must have been colourful to say the least.

So, other than moving Alice to Shepherd Bush, siring a couple of sons and establishing a name that would become part of a well-known catering double-act eighty years later, what did Wilfred add to the concoction? Gus, without doubt, inherited his forcefulness, business acumen and drive from Alice; but the charm and the boulevardier, that certainly didn't come from his mother so, probably, was pure Wilfred, although I'm sure Gus would have denied it.

Gus was christened Marshall Augustus Payne, but he

was never known as anything but Gus. The name Marshall came from Wilfred Payne's older brother, who drove a Black Maria for the Metropolitan Police. There was an amusing time years later during the war, when Gus was a Warrant Officer in the Royal Engineers and he received a letter addressed to 'Marshall Regimental Quarter-Master Sergeant Augustus Payne', which sent the whole of Longmore Camp into paroxysms of laughter wondering how he had acquired this new rank.

Consequently, there he was, with a hard working ambitious mother and a lazy, gallivanting father. To add to the mix he had had an elder brother with the unusual name of Greenie who was very much Alice's favourite, but sadly he died at the age of seven, when Gus was only five. For whatever reason, thereafter Alice took totally against Gus, and indeed all boys, which certainly was to affect me later.

What home life Gus had as a child was with Wilfred's mother, Gran, who, I have been told, was also a strong personality – the only picture I have of her is a verbal one from my mother who said that she had no teeth and so spent the whole of Sunday afternoons sucking a piece of pork crackling – sadly she died during the war in the middle of the Blitz, a few years before I was born.

Gus's childhood was spent almost totally on the streets of Shepherds Bush, behind the stalls of the market, with lunch at his mother's coffee stall and, his salvation, swimming at the Lime Grove Baths.

The dominant part of his young life was the Penguin Swimming Club and he was very competitive but a team player, particularly in water polo. He swam for the club, for London and for the southern counties but never at international level. It was a very important ingredient while growing up with no home support and minimal education.

He left school at thirteen and started working for the largest local employer, J Lyons & Co, a mile away at Cadby Hall in Hammersmith Road. Lyons was a model employer, a very large private catering and food manufacturing company created by two far-sighted and paternalistic Jewish families at the end of the nineteenth century.

Ironically, considering how his later life was to develop, he never had any deep involvement in the catering side, mostly the bakery and distribution. Over the next fourteen years, before he went off to the army in 1939, he became one of the up-and-coming, non-family members of the management team.

His promotion was in spite of his lack of education but was doubtless helped by the time he spent, and the leadership he demonstrated, at the very large Lyons Sports Club at Sudbury. The club, with one of the country's top tennis teams, of which Gus was a leading light, was, along with Shell's Club at Lensbury, a model of the paternalism of the time.

While most of the Lyons management went into the Catering Corps at the outbreak of war, Gus joined the Royal Engineers, Movement Control – or Muddle and Confusion, as he called it - rising to RQMS. He was shipped out to France with the Expeditionary Force and returned from St Nazaire two weeks after Dunkirk and the ship he was on picked up 2000 of the 5000 troops from the accompanying ship, the SS Lancastria, when it was sunk off the French coast by German bombers. The only news my mother heard was by a telegram from Plymouth: 'In England don't worry - Gus'. The rest of his war was spent in England, preparing for D-Day and beyond, and was not particularly notable militarily, although I'm sure that whatever unit he was in had as good a time as any in the British Army.

Before carrying on with Gus's story, with him setting up his own business having, like all his generation, lost seven years to the war, it is time to introduce the driving force of his life, my mother, Lila.

Lila was a very beautiful lady, five foot eight, (the same height as Gus), slim and elegant, she was often mistaken for Lady Isobel Barnett, a television personality of the late fifties, and she and Gus were, without question, soul mates. Gus was everything for Lila and he would have done anything for her. They met, in a dance hall in Shepherds Bush, both of them seventeen years of age, in 1928 and were married in 1936, when they could afford it.

Lila knew that Gus had been responsible for taking her out of a very grim existence and Gus's absolute motivation was in ensuring she never had to return to it and helping her to enjoy the good things she had been denied.

The story of her childhood has only recently been pieced together, in part by me but largely by my sister Janette, following conversations that both of us had with Lila during her final illness and also in talks with our 'Auntie', Marie Tugwell, who was my mother's best friend at school.

Lila's mother - another Alice - Alice Freeman, was the daughter of a comfortably off Jewish family, who met and married James Turner around the turn of the century. His family, also well off but Christian, and the Freemans, totally ostracised James and Alice and gave them no subsequent help.

James was invalided out of the First World War with tuberculosis and died soon afterwards, leaving Alice with no means of support and four children, one retarded through meningitis, having already lost two. Alice had an operation which went wrong, leaving her for the last twenty years of her life with open wounds on her stomach, and no way of paying for medical help.

When her eldest sister died, also of TB, Lila and her elder sister Vi, together with their retarded brother, were sent to Dr Barnardo's for some time, but returned home to look after their mother, including tending her wounds on a daily basis.

One of my greatest regrets is that I never met my grandmother, Alice Turner, who died before I was born, because from everything I've heard and all that I know of Lila and Vi, she must have been very special. She was an exceptionally proud woman who, even in total penury and in constant and considerable pain, established a set of standards of behaviour for her family that were absolute.

My 'Auntie' Marie told us that the Turners were always considered 'a cut above', although no one, not even Marie, was ever allowed past their front door to witness the fact that all their furniture had been sold to keep them going.

When all this story came to light just before my mother died at the beginning of the nineties, it cast a new ironical twist to the subsequent story of Lila's sister Vi, who met and married Phil Kushner in the middle of the nineteen-thirties, prompting his wealthy Jewish family to disinherit him for marrying 'out'. None of them knew, since Vi had been brought up a Christian, that by Jewish law, since her mother was Jewish, so was she. Families are weird, religions even weirder.

So this 'Penny Dreadful' story was what Gus took Lila away from and, once the war was over, he determined to ensure that his life was dedicated to Lila's, and his own, enjoyment.

Alice Payne, meanwhile, had sold her stall and moved into a café in the market and among her regulars, chalking up their lunches on 'the slate', was a young Jack Cohen, founder of Tesco, who once or twice a week, would set up a stall selling cans that had lost their labels: "Come on, take a chance Lady!"

The café was bombed and she moved to very much larger premises at 204 Uxbridge Road, which remained the home of the family business until the end of the seventies.

The problem with having the larger café was that it meant employing more people and Alice was, without question, an appalling employer. Every week she used to sack one or two and Gus, who was still in the army based at Longmore Camp in Hampshire, had to come up to Shepherds Bush each weekend and try to persuade the staff to come back to work on the Monday.

He left the army and returned to work at Lyons, but still helped Alice at weekends. One day he was asked if he could do a Home Guard Reunion Dinner in a local church hall; he agreed, enjoyed it, and his life was changed forever.

He started to do more and more weekend parties while still working for Lyons, culminating in the wedding of his doctor's daughter in Chiswick. Dr Wilf Winder, who was to remain my doctor until I got married, was married to the sister of our family's solicitor and so Gus felt honour bound to do their party, the social event of the year in Bedford Park where Gus and Lila had just bought a house, although he was ill equipped to do it.

As Gus told the story, he took seventeen journeys in his Hillman to get everything from the Shepherds Bush café to the lovely, and very grand, Norman Shaw house on The Avenue in Bedford Park where the Winders lived and the wedding reception was to take place.

In order to reduce the journeys coming back, Gus took out the rear seat and, sure enough, it only required fifteen trips. Sadly, someone stole the seat from outside the Winder's house and so that was the profit for the job gone. Gus's first away venture was not an auspicious start.

3

ELEVENSES AND JUST DESSERTS

So followed more weekend functions for Gus and every other week during the winter the home game at Queens Park Rangers meant the serving of eight hundred egg and bacon breakfasts. As Gus told me, selling these for one shilling and sixpence each when they cost tuppence-farthing, made him wealthier than at any time for the next decade and a half, and forced him to decide whether he could keep on doing it on a moonlighting basis.

He resolved to burn his bridges at J Lyons & Co and start his own business full-time. His boss at Lyons, Sam Salmon, told him: 'Payne you're a bloody idiot!' Gus saw him many times in the years ahead and the, by then Sir Sam, nobly agreed that he always knew Gus was going to succeed and had only said that to persuade him to stay.

Gus started in earnest as an outdoor caterer in 1947, taking on the lease of the Old Fire Station in Pennard Road as a warehouse in addition to the café at 204 Uxbridge Road.

It was, of course, the most monstrous time to start trading as during the war the country had had seven years of micro-management by the government, by necessity, given what we were up against. However, this left a long-lasting residue of heavy-handed bureaucracy and an atmosphere of envy that did not allow or encourage business to flourish. Unlike almost any other country involved in the war, Britain still had rationing a decade after it had finished, shortages of almost everything, and the country was crippled by archaic labour relations and practices. Altogether this caused all our immediate neighbours, including those we had beaten in the war, to leave us behind for the next thirty years.

Writing this in the new millennium, I have to keep readjusting my thoughts to the prevailing attitudes and circumstances of forty years ago. Having lived for the last twenty years with the relatively benign tax regime that graciously allows us to keep sixty percent of what we earn, it is very difficult to remember that governments were daft enough to believe that they could rely on the most successful of their citizens to hand over up to ninety-eight percent of their income without demur.

Looking back, it is quite astonishing that there were no tax riots, as there were for the vastly less draconian Poll Tax although, in reality, in the 50's and 60's, those who were the most affected were also the most effective at dreaming up ways of defeating the system, without recourse to violence.

For thirty years after the war, a vast part of the ingenuity of the people of Britain was diverted to this task. Quite simply, by far the largest industry in the country during that time was people stealing their own money, culminating in the ludicrous scenario of major companies owning the suits that their directors and managers stood up in.

Since Gus grew up after the First World War in Shepherds Bush Market, he was not unversed in the dark arts of the black economy. The denizens of the world he knew as a child had two currencies, cash and barter, and when, much later in life, he was working in very different surroundings, his instincts remained true to these. Even though by the time I joined him he was dealing on occasions with some of the biggest businesses in the land, these two currencies stood him in good stead as all and sundry wanted their income, and more especially their expenditure, unrecorded.

In those days, business entertainment was seriously frowned upon by officialdom and could be costly in tax terms to both the host and guest. It was soon discovered by the sponsors of golf tournaments that they could lose the payment for the entertainment of their clients within their totally acceptable sponsorship bill. Those clients also enjoyed a day at which they would be surrounded by visual reminders of who their hosts were, and be wined and dined in some considerable style.

At that time the only entertaining at these events was by the sponsors themselves, looking after (not necessarily in order of their priority) the players, the press, the sponsor's clients, and their own directors and senior managers. This was where Gus made his mark. While looking after a growing band of golf sponsors, including Daks, Dunlop, Martini, Penfold, Rothmans, Gallagher, Schweppes, Bowmaker Finance as well as Teachers and Ballantine Whiskies, Gus became a considerable feature on the golf scene and in his inimitable way made a lot of very important and interesting friends.

Gus decided, and I'm not sure at which event first, that he would get the tentage contractor to put up a small marquee, thirty foot by twenty, attached to the main kitchen, where he would be able to entertain his own friends and existing and potential customers.

This was the origin of the corporate hospitality

pavilions for various companies unconnected with the sponsor that has become ubiquitous at sporting occasions and a major source of funds for virtually all events.

Since Gus enjoyed drinking Champagne at eleven o'clock in the morning, and only at eleven, he always changed from his morning preparation garb into Burlington Bertie complete with carnation, in time to enjoy this, along with any of those in the know who happened to be around and could find his tent. After a few years 'those in the know' became quite numerous and increasingly adept at finding his hidey-hole. Thus commenced the instant tradition on the golf circuit of elevenses, which in latter years would make the Bollinger Tent *the* meeting place at the Open Championship.

Gus's hideaway at the golf events had, since its inception, been looked after by Connie Rouse, Gus's Girl Friday and senior waitress. Connie had met Gus in the Army during the war and had joined him as soon as he started the business. She was the original cockney sparrow, a dark haired Barbara Windsor.

She lived with a team of brothers and her long-term Glaswegian boyfriend, 'My Jim', in a London County Council house in East Acton. I don't think anyone ever discovered why Connie and My Jim never married, which was unusual in those days.

Connie was totally besotted with Gus, although there was never a hint of romance, and working for him was the most important thing in her life; given the hours she worked, it was just as well. She was a tireless support for Gus and at events she would be the driving force, pulling together all the myriad things that come together in catering as 'mise-en-place'. She also had a lovely line in malapropisms that kept us in fits, as on the occasion when My Jim was starting driving lessons and she said that she was going to the post office "to get him a provincial licence".

When Gus started having his own guests at these events, it was a natural for them to be looked after by Connie, quite often as a total side issue to her main job which might be looking after the Players & Press Marquee or preparing the starters and desserts in the kitchen. She was the person for whom the current buzzword 'multi-tasking' was coined.

Very soon the guests knew Connie as well as they knew Gus. She looked after them extraordinarily well, remembered what everyone drank, names of their wives and children and never left anyone waiting. Quite an outstanding person, and it was she who was to be my first mentor in the very special art of hospitality. I little knew when starting to learn these skills with Connie that within a few months I would not only run into Her Majesty The Queen for the first time, quite literally, but also discover just how embarrassing this catering business could be.

The first summer season of 1961, after the initial outing at Bournemouth Tennis, consisted of ten golf tournaments, eleven agricultural shows, ten company sports days, eight factory open days and a week at Farnborough Air Show.

No two events were alike but usually I would start off as van boy, unloading equipment and helping to set everything up. This was very heavy physical work and for most of the major events took anything up to two days to complete.

On the first morning of an event, we would start at about six o'clock and be working on preparations until ten thirty. At this point it was time to get changed and for the first few years this would entail me dressing in waiter's garb, black trousers, highly polished shoes (inspected and passed as suitable by Regimental Quartermaster Sergeant Gus Payne), white shirt, black bow tie and short, white mess jacket. If it was an event where Gus had a hideaway, it was often decided that I should work in there with Connie. It was the best possible apprenticeship that I could have had.

Initially I was just Connie's commis, in other words I did the fetching and carrying for her. She taught me the value of setting up everything that might be needed for the entire lunch period before anyone arrived. She was always so particular about this, although, annoyingly, she was also very fond of saying "I always say it's the little things that take the time" several times, every single day.

Gus's hideaway was, invariably, a 30-foot long, 20-foot wide marquee, usually with green and white striped lining. There was no entrance doorway, merely a flap opening in the narrow end furthest from the connection to the kitchen marquee. The reason for the flap was so that no one would come into the tent by mistake – you had to know it was there.

The tent had a bar area and trestle tables and a couple of dozen fold-flat wooden chairs. A corridor led through to the kitchen with side tables containing all the back up and re-lays for the lunch.

We would usually know beforehand of about a dozen or so people who had been officially invited, with passes sent. On top of that there were another dozen uninvited but expected guests, followed almost invariably by another couple of dozen uninvited and unexpected guests. It was therefore not uncommon for Gus's Tent to serve forty to fifty people a three course lunch – absolutely free - as well as various people just calling in for a drink, and this was after the twenty or so of 'those in the know' had had their Champagne Elevenses.

Until my arrival on the scene, Connie had traditionally looked after this lot on her own; now she had me as a runner, which allowed her to undertake more jobs in other parts of the event. Within a few tournaments, this had progressively changed as Connie moved to other duties and I was left in control of Gus's domain.

I say in control, but the fact was that if Connie hadn't kept a very acute eye on me, I would have been badly floundering. It did however give me a very powerful insight into working under intense pressure, maintaining a high standard and never letting the guest know that you were absolutely terrified, pretty well out of control and within an ace of total panic. Since the guests were Gus's personal friends and some very important clients, how I dealt with them was critical.

It was a lesson that stood me in good stead for the next forty years. Not the art of being in control – the art of *appearing* to be in control. The emblem of the caterer should of course be the swan – gliding serenely on top of the water and paddling like fury underneath.

Gus decided that, after the first few events, I needed to be trained outside the confines of his company. What he really wanted was for me to be 'blooded' in the more difficult aspects of outdoor catering and he was well aware that his own staff were not going to tell his son how they were cheating his father; I was going to have to discover that for myself by working for some of his rivals.

His two main, much larger, competitors at that time (as they have been, in different forms and ownerships, for my entire catering career) were Letheby & Christopher and Ring & Brymer. Both were old-established companies who, to a large extent, shared the catering at the major racecourses, where Gus went, but always as a punter, very rarely as a caterer.

L&C was run by a delightful, urbane, gentle man by

the name of Colonel Len Livingstone-Learmonth, every inch the tall elegant country squire and known for fifty years or more by the Queen Mother as Len. R & B was in the hands of Lindsay Ring, also a tall dignified, but much stouter, character and the latest in a long line of his family in the business. For one year they decided to merge - not very successfully.

It was to this uncomfortable merged entity of Lethebys and Rings that I was sent to work for the madcap week of Royal Ascot less than two months after joining Gus. This was following the feudal tradition of sons being trained by competing family businesses or neighbouring estates.

I drove to Ascot on the NSU scooter that had been handed down to me by Gus. He had purchased it five years earlier at the time of the fuel shortages following the Suez crisis, when he drove round the West End of London wearing a bowler hat rather than a helmet. I parked in the massive staff car park half a mile from the racecourse. Since Ascot was considerably bigger and vastly more complicated than anything Paynes did at that time, and being only two months into the job, I found the entire operation baffling and very daunting.

It was also an operation that had been going for a hundred years and some very sinister Old Spanish Practices had grown up. It was in fact a seriously corrupt place and Gus was right to send me there, as I learnt a tremendous amount about the disreputable ways of our idiosyncratic industry.

I discovered that quite a number of staff who regularly worked for Gus were at Ascot, but were happy to go there on half the wages they got from him. In fact there were some teams who came down each day from Yorkshire and Liverpool for the same amount. I realised they must have had some reason to do so.

That particular week I was given the job of looking after one box of eight people in the brand new grandstand, which was opened that year. My client was a racehorse owner, and friend of Gus, naturally, but that didn't stop him giving me a very hard time, keeping me on the go all the time with last-minute requests and giving me hell if he, or his guests, were kept waiting.

My observations of the fiddles that week were limited to service in the boxes and restaurants but these alone were legion. Waiting staff were given tips by their customers, not just for the service they gave, but also for the extra items they were able to supply without them being officially charged by the caterers. This applied to restaurant hosts as well as box holders.

The waiting staff had to distribute some of the tips down to the kitchen or cellar staff, in order to get their illicit supplies, and I know that a great many of the floor managers were also tipped to turn a blind eye. The cancer was therefore endemic in the entire catering body.

There was a complete bartering system in place. For example a member of staff would corner one particular commodity – such as celery salt – and trade it at the appropriate time – when gulls' eggs were on the menu – for something else that their host wanted. In later years, working at Ascot, I rather enjoyed taking part in this bizarre bazaar, but at sixteen and very new to the trade, I found it really quite frightening.

I am well aware that I was very naive but I found it appalling and couldn't understand why, by paying such ridiculously low wages, the companies seemed to be colluding in the roguery. It was, of course, all to do with the universal belief that it was good to leave a few fiddles in the system so that everyone could avoid paying ludicrous taxes. I worked at Ascot every Royal meeting for the next ten years and it served as an annual update on the tricks of the trade.

This first year, as a total novice and unversed in the ways of backhanders, one of my main problems was making sure that I was able to deliver what my client expected; it was quite a battle, which I just about got through. This naivety did however cause one almighty blunder on the first operational morning.

Around eleven o'clock, just before my host and his guests were due to arrive, I discovered that the cellar had not delivered any Pimms to my box on the fourth floor, because I had not given a bung to the appropriate person. I had to go down to collect a couple of bottles before my client arrived. Members of staff were allowed to use the lift up to ten-thirty, but not after that, so I knew I was taking a chance, but after all it was five floors down – and up again.

So I pressed the button and as the lift arrived and the door opened, I jumped in and very nearly knocked over the Queen, who was being shown round the new grandstand by the Duke of Norfolk and the boss of the caterers, Colonel Len Livingstone-Learmonth. Her Majesty smiled charmingly, realising my mistake, the Colonel, recognising me as Gus's son, tried to stifle a chuckle, as he looked at me with his almost-closed crinkly blue eyes, but the Duke's displeasure was all too evident by a glare that could have frozen Hades.

As well as problems with supplies, I was seriously deficient in service skills having had an intense but limited training with Connie. I also had problems with the old-fashioned waiter's uniform I had to wear. It was a heavy, serge, black tailcoat which was very hot, under which I had a starched white shirt with separate stiff wing collar that stuck up into the underside of my chin.

I had just started shaving and my sore neck suffered from unaccustomed razor-burn, aggravated by the collar, which meant that I found it very difficult to look down. Together these facts contrived to create quite the most embarrassing moment of my young life, which could easily have made me give up catering totally.

One of the guests on the second day, was a very attractive girl, possibly a couple of years older than me chronologically, but decades older in sophistication, wearing her Ascot best, an off-the-shoulder dress that revealed more than it hid. And what it revealed was, or rather were, to an innocent lad of sixteen, virtually unknown and seriously lovely.

Connie had taught me well about silver-service, the process of using a tablespoon and fork to serve dextrously from a salver onto the customer's plate, so I made it through the first and second courses without mishap. Then came dessert.

This consisted of a large silver tray full of individual crème caramels. Silver-serving crème caramels is not dissimilar to catching hold of a bar of soap in the bath. It is just about achievable when there are two or more on the dish as you can chase one with your spoon and fork and push it up against a second one to catch hold and serve it.

With the last one, of course, you have nothing to stop it against and it keeps running away from you around the silver dish. This, coupled with an inability to glance down, due to a razor-sharp collar, is seriously hazardous. All that is then needed is a luscious young sophisticate that you have been secretly lusting after as the person who is due to be on the receiving end of this final crème caramel and you have a scenario from hell – with the predictable outcome. The final crème caramel had enough of running away from the spoon and fork and made a break for freedom and suddenly the lovely dress had three not two peering out of its top.

Squeals, shrieks and massive embarrassment follows, and that's just me. The fact that everyone, including the object of my lust and fumbling incompetence, finds the whole thing hilarious does not help at all. My mind has drawn a blank over what happened next but I am sure the word aplomb was not one that came anywhere near. Inevitably the story got round Ascot like wildfire and the consensus among the staff was that, once the crème caramel had leapt away, I should have breathed on the spoon to warm it and gone in after it.

I'm glad I didn't know that at the time!

4

To All In Tents

By the time I joined the company, Gus had become a very close friend of the many delightful people who ran the Dunlop Sports Company, most especially Eric Hays, their golf manager, slight of stature, seemingly always dressed in tweed, with a deep Scottish burr and with a somewhat other-worldly innocence about him. Over the previous twenty years Eric had become the supremo at running golf tournaments, and in particular the Dunlop Masters. This was the largest sponsored event in the country at the start of the sixties.

Dunlop Sports were not only the biggest suppliers to the world of golf and tennis, owning the Slazenger brand as well as their own, they were also part of a massive world-wide industrial conglomerate with large subsidiaries in, among other things, tyres, chemicals and engineering. They were also one of the first companies to recognise the value in using their own event, run by their sports division, to entertain customers from all the other divisions, each of which paid towards the cost of running the biggest and most prestigious golf event after the Open Championship.

Gus had already looked after the catering for half a dozen of these tournaments, each one larger than its predecessor, as the Dunlop divisions increasingly recognised how effective this was as a way of promoting their individual businesses very enjoyably.

The first one I went to was towards the end of my first season and took place at Royal Porthcawl Golf Club on the South Wales coast. In those days, before motorways and Severn bridges, getting there required a fearfully long journey through Oxford, Gloucester and Monmouth crawling up and then racing down hills in the dreaded Bedford 'Custard-Powder' wagon. At least I was travelling with little Tommy Potter, who kept me enthralled with tales of his adventures with the Eighth Army in the North Africa campaign in World War II. He saw this all the way through, from being driven out by the Italians and then slinging them out, being pushed out by the Germans and finally defeating Rommel at El Alamein. He gave no impression of being a hero as one of the famous 'Desert Rats'; indeed he quite happily put up with the joshing when, because of his size, we referred to him as a 'Desert Mouse', but he certainly had some hair-raising stories to tell.

Royal Porthcawl Golf Club is at the end of a long, winding road out of the town, all along the cliff-top, the edge of the road being marked by large white stones every yard or so. Before the clubhouse comes into view on the bleak headland, you first see its neighbour, a forbidding Gothic monstrosity, straight out of a Hammer Horror Movie, which is, or was, the Miners' Rest Home – guaranteed to make you want to go back down the pit as soon as possible.

As we drew up in the old van, a rather harassed young man in blazer and flannels, about six foot tall with formally parted black hair and a considerable air of resolution, blocked our way and wanted to make sure we were authorised to come in. When that was established he informed us that we must keep to the paths and not spoil the grass areas between the big marquees. This was my first meeting with Richard Brown, Eric Hays' assistant and future successor, who was to be an important figure in my life and is still a great friend nearly fifty years later (I have to say that during all that time, Richard's prime concern has always been to make sure that no one spoiled the grass area between the big marquees!).

This was his first tournament and his first introduction to Gus who was to bamboozle and charm him for the next couple of decades. That Gus took to him was indicated by the fact that he was referred to by Gus as his 'number two son', in the style of Charlie Chan. For all that the future held for him as a big operator in the world of golf, at this, his first event, the main task given to him by Eric Hays was to look after the ladies' loos.

It was a seriously windswept course and the old-fashioned ridgepole tents flapped like fury, but I found it very exciting because it was so big and busy, and this in spite of the fact that, like Richard's, my duties were less than inspiring. My first job each day was to wash out the galvanised metal dustbins we used to store our potatoes. There were, of course, no sophisticated suppliers of ready-prepared vegetables in those days. Part of the preparations for any event was to strike a deal with a local fish and chip shop to have them peel however many hundredweight of potatoes we were going to need each day. We then collected the dustbins we gave them to put the potatoes in; a dustbin full of potatoes and water took some lifting on to the back of the van and off again on site – no tail lifts on the vans either.

Writing earlier about having to remember the prevailing business attitude forty years ago, I am now reminded that there was another almighty difference to running a business then and now – the changes in equipment and the lack of support systems and sub-contractors.

In the same way that any young person today watching a re-run on TV of an nineteen-sixties' episode of 'The Avengers' (on what my wife referred to as the Sad Gits' Channel) must ask themselves why it is that Emma Peel doesn't inform her suave, debonair partner, Steed, that she is held prisoner in the clutches of a lunatic professor in the bowels of a castle in deepest Kent, by simply texting or phoning him on her mobile, I have to hark back to the fact that these were days when the equipment available to us was very different from today.

Bear in mind that the British Army was, at that time, still using the same basic field oven designed by the great chef Alexis Soyer for use in the Crimean War in the middle of the nineteenth century. We, lacking the cranes and heavy transport they had, made do with making our own ovens, or rather hot-cupboards, from sheets of aluminium around a light steel frame containing a Calor gas ring. These would sit on trestle tables, which had a sheet of aluminium tacked on the top to save it catching fire.

These metal-top tables were made by Gus's staff during the winter, as were all the grey-painted wooden boxes carrying the thousands of different plates, cups, cutlery and a thousand and one assorted bits of equipment. A whole array of such tables with hot cupboards as ovens and Calor gas rings in steel frames as cooking ranges was the ramshackle basis for a series of kitchens that could serve several thousand people a day with hot meals. Heath Robinson it may have been, but it worked - usually. The

Calor gas appliances, particularly the hot cupboards, very often blew back and, having more than once lost the hairs on my arms, I approached them with the same trepidation that one might have for Arkwright's till. On many occasions they caught light at the rubber connection or had a build-up of gas that caused serious explosions.

One time a few years later at the Gevacolor Tournament at Stoke Poges, where we had two kitchens a few hundred yards apart, I went into one kitchen, having got dressed ready for the lunch service, to discover Jimmy Kirkham, the Liverpudlian chef who was a director of the company, frantically flapping at flames on the top of his head. I rushed him to hospital, only to find, to my horror that Pip Thow, Jimmy's junior, was already waiting for treatment, having had the same thing happen to him in the other kitchen. I left them both there and hurried back to assume the unaccustomed role of chef for the lunch that was about to be served. Two hours later after a very nerve-wracking lunch service, which proved to me that I was quite right to stay in the front of the house, the two injured chefs returned together swathed in bandages but otherwise none the worse for their ordeals. Amazingly, we operated at that time without any refrigeration. Our first refrigerated van was tailor-made for us in the late sixties and, when we replaced in the early seventies it was sold on to our larger rivals, Letheby & Christopher, being their first fridge van.

We never, to my knowledge, poisoned anyone or served anyone a bad or off meal - the differences in the approach of the environmental health authorities could explain why.

When I started in the business, all the hygiene rules began with the assumption that we caterers bought pure food and all the rules we had to follow were to ensure that we didn't subsequently contaminate it. By the time I stopped catering in the late nineties, the assumption in law was that we, the caterers, were being supplied with poisoned food and it was our job to correct that by the cooking process. In other words killing the poison by over-cooking. Not all the changes in forty years have been for the better!

At the Masters at Porthcawl, in three days we served nine thousand lunches to the ordinary customers of Dunlop and a further nine hundred to the divisional bosses and their VIP guests in separate tents. Since this was a very busy affair, Gus did not have a hideaway at this event; in any case most of his normal visitors would naturally have been lunching as guests of Dunlop Sports. I was delegated to help Alan Cox manage the seating in the main marquee. This seated eight hundred and we served two and a half thousand on the Thursday and again on the Friday. On the Saturday, which in those days was the last day, when the golfers did two rounds, we served over four thousand. In other words we did five sittings; given the fact that some people took longer than others, some of the tables were used six or seven times.

We were working to a queue, with Big Alan standing by the control desk sweet-talking the waiting punters and indicating to me with his fingers how many seats the next party needed and I had to hustle the existing diners to make space. I then indicated to him what was available and showed the new guests to their places. We also had to make sure that the hundred or so staff in the tent were doing their jobs properly, as well as dealing with the customers' problems, and this went on from midday to four by which time we had to reopen for afternoon tea. The queue was anything up to two hundred yards long. It was seriously daunting and very hard work but was an object lesson in charming the customers to do what we wanted them to do.

What was even more amazing was that we, the juniors, used to sleep in this massive tent, pitching our camp beds between the tables; in the middle of the night, with no lights, it was very eerie. The first couple of nights, after several days of doing the very hard physical job of setting up, we went to bed straight after supper and slept the sleep of the truly exhausted. However, after the first day of the tournament I was so hyped-up with the excitement and the nervous energy I had used that I had to go out with the team to the local pub before collapsing once again in my bed to have a few hours' sleep before Jimmy Kirkham, the chef, as was his habit, brought us a cup of tea at 5.30 to start the day.

The team of staff, both full-time and regular casuals, almost invariably went out for the evening as a group but I had decided very early on that I would not make a regular habit of going out with them.

Firstly, having spent the day with a rictus smile on my face, in the company of large quantities of people I had to charm, I rather enjoyed a quiet time on my own, sometimes just reading. Occasionally I spent time just looking round the lovely towns we were visiting with the help of my new best friends, Nikolaus Pevsner and his books, Buildings of England to whom I had been introduced during the few months of studying architecture in the sixth form at Latymer.

So, on the Friday night, since I couldn't at that time drive, I got a lift into town but left the others to carry on to their pub-crawl. My thoughts that evening were not quite so cerebral as to embrace Pevsner; I went to the local Palais-de-Dance in Porthcawl, which, strangely, was not mentioned in his book. I struck lucky and met a very nice girl called Gill, whom I escorted home after the dance finished. I considered I was even luckier when she invited me in as her parents were away for the night. I blush today at the sheer innocence of it. We sat on the sofa and chatted. She told me her dad was not interested in golf as his sport, like the majority down there, was rugby. He was a second-row forward for the local team and had played for Wales.

We just about managed a brief cuddle while listening to Gene Vincent and Connie Francis records on her Dansette record player when there was a sound of movement outside and she squealed in her lovely Welsh lilt, "It's my parents, you've got to go. My dad'll go mad." Having no desire to meet a maddened Welsh second-row forward at two o'clock in the morning, I needed no further prompting but skedaddled out of the back door, suddenly realising I had a very long walk back to the golf club.

What I hadn't appreciated on the previous occasions I was driven up the road to Royal Porthcawl was the fact that there were absolutely no lights and as this was a new moon, the road was in total, total darkness, once I had left the town. Trying to remember the road, I had this vision of the cliff along the left hand side and was terrified that I was going to fall over. To be safe, once I got further up the hill where I thought the cliff began, I started crawling along on my hands and knees, feeling for the stone markers at the side of the road. It took a hell of a long time but then I saw when outline of the Miners' Rest Home looming on the skyline. Suddenly there also appeared to be a moving light, very faint but growing brighter. It appeared to be getting closer faster than I was approaching the home. It also appeared to by swaying from side to side. Slowly it dawned on me that it was a bicycle lamp. Sure enough it was the local copper, on his rounds who, save for a cheery "Good Morning", totally ignored this crawling young man on the cliff-top road in the wee small hours. Obviously a common occurrence in Porthcawl.

I made my way eventually to the massive marquee and my camp bed, which fortunately had not been 'got at' by my colleagues – a habitual hazard of a peripatetic catering life – and a couple of hours later Jimmy woke me with my tea to face the busiest day of the week. Four thousand lunches, four thousand teas and then we packed everything up ready for the return to London on the Sunday.

Having successfully negotiated this logistical nightmare and crammed everything onto the vans, we just had time for some fish and chips and a pint of beer before the pubs closed in Porthcawl. A last night sleeping in the now totally empty marquees and we then, on the Sunday morning, handed over the scene of desolation that is an event site after a tournament to the tentage contractors for them to dismantle and erect at the next event. Then the whole shebang of cars and vans set off on the tortuous route back to Shepherds Bush led by Gus in his Humber with Connie and a couple of waitresses. They obviously travelled faster than us, but Gus decided that the whole team deserved a good lunch after the week we had had, so he stopped in Chepstow, a couple of miles before the English border, and booked twenty of us in for lunch at the Beaufort Arms Hotel. He then, in the style of Phineas T Barnum, waited in the town centre and waved us all in to park and ushered us all into the hotel.

We all went in to the hotel restaurant en bloc; rather a noisy procedure as was noted on the dour faces of the diners already having their lunch, dressed in their black Welsh Sunday Best. Once we had settled down Gus asked everyone what they wanted to drink and gave the considerable list to the waiter who informed him that this was Sunday, Chepstow was in Wales, just, and there was no drinking in Wales on a Sunday. Guffaws of laughter from Gus at his two-mile blunder and we all got up en bloc and departed for an English pub. It was now very obvious why the diners had been so po-faced, not being allowed a drink.

Dunlop continued to be our largest client, and the Masters one of the largest events, for the next two decades. Traditionally it moved each year to a different golf course and the year after Porthcawl it went to Wentworth in Surrey, one of the grandest clubs in the Home Counties, surrounded by a very exclusive residential estate. In order to accommodate such a large event, Wentworth had to grub up the very fine double row of box hedges that had stretched a couple of hundred yards away from the front of the clubhouse. So extensive was the tentage for the ever-growing event that Henry Longhurst, the doyen of golf scribes and first television commentator, referred to it in his Sunday Times column as 'what may veritably be described as a tented village', the first time that frequently repeated phrase was used.

Eventually the Masters became too big for all the guests Dunlop wished to entertain to be treated to a full lunch. It was decided that in order to allow the numbers to grow, the also-rans would now be given a lunch box and drink and only the VIPs would have a lunch. The obvious happened and the list of VIPs trebled, but that still meant many thousand lunch-boxes had to be created and packed on site.

By the mid sixties, when the tournament was located at Little Aston Golf Club, which is near Birmingham and therefore the heartland of Dunlop Tyres, the largest division and the biggest paymaster of the event, the numbers had grown to five thousand VIP's and some sixteen thousand lunch-boxes.

Because a great many guests came more than once to each event, the menu in the lunch-boxes was changed once during the week. One item proposed to be included in the pack was a scotch egg, which Gus, against the advice of Jimmy Kirkham and myself, had agreed would be provided by Andy Cotterill, an old rogue of his acquaintance who ran an airline catering business. Gus, on my prompting, had phoned Andy a couple of days beforehand to check the delivery and was told that everything was under control. Quite patently this call had reminded him of the need to make ten thousand scotch eggs and get them delivered to Little Aston.

Sure enough, Andy the large, somewhat scruffy and seriously sweating airline caterer arrived, very late, in his Bentley, the seats of which had been removed – shades of Gus and his doctor's wedding in Chiswick – and boxes of scotch eggs, still a bit warm, each wrapped in cellophane, had been forced into the boot and interior of the car. We got the very squashed eggs out and unpacked them and, many hours later than planned, started the production line packing the lunch-boxes for the following day.

By midnight we had about three-quarters of the boxes done and decided to call it a day. I got into my bed in the next tent and quickly fell asleep. I have a very keen sense of smell and I am sure that I would be woken by smoke in the event of a fire. This night I was woken after a couple of hours by the unmistakable smell of rotten eggs – ten thousand of them. I got up and looked at a couple and knew the worst. I was walking over to where Jimmy Kirkham, the chef, was sleeping in a caravan to discover him already walking over to the lunch-box preparation tent. He hadn't been able to sleep thinking about the bloody eggs. We set to, making a batch of tea before waking everyone up and starting the process of unpacking all the eggs. I contacted one of our suppliers of pies at his home, waking him up, and asked if he could get five thousand sausage rolls to us by eleven o'clock and another five thousand by two. He did, and, after an enormous amount of hard work, the Dunlop customers, oblivious of the hiatus, were satisfied.

We had, however a remaining problem – what to do with the ten thousand rotten scotch eggs in their cellophane bags. We didn't want to put them in our rubbish skips, as we did not want our clients to know of the problem. It was decided that they would be put into the massive soak-away pit we dug for the washing up water outside the back of the tent, on the assumption that they would decompose and rot away. However I did not check that the boys who were instructed to get rid of them had, as required, taken them out of their wrappings. They didn't and the scotch eggs in cellophane formed an impervious layer at the bottom of the pit and months later the farmer, whose field Dunlop had borrowed to pitch their tents, was totally bemused as to why the water had not drained away and he was not able to take down the surrounding fence and let his animals back into the field.

At this particular event one of Gus's particular friends, Henry Longhurst, doyen of golf writers, shocked everyone by arriving in a brand-new American Ford Mustang convertible, which didn't seem to go with his image as a portly, jowly, eccentric Englishman, dressed in very shabby tweeds and almost invariably 'in wine'. Henry and Gus were the same age and shape and delighted in each other's company. Both liked the bottle. The difference was that Gus was always supplying the drink for everyone else to consume, usually at his own expense. Henry did not believe in buying drinks. He believed that everyone should be quite happy to supply him with drinks in order to hear his stories. I don't believe that anyone would have argued. I was actually present on the only occasion when Henry ordered a round of drinks – at the Saunton Sands Hotel on the evening before the Schweppes PGA Championship in April 1966; it caused total consternation - until someone else stepped in and paid for it.

The reason why I specifically remember the date is that it was the day after one of the most important days in my life – when I proposed to Judy – just before I left to come down to the first tournament of the season. At the end of the first day setting up the event and just before I went over to the hotel for a drink with Gus, I received a call from Judy telling me that she had been to Burlington Arcade with a friend and chosen and bought a ring during her lunch hour. It was Gus telling Henry about my engagement that prompted the unaccustomed burst of generosity that made him order a round of drinks. Fortunately the burst passed before the time came to pay and Henry's reputation was saved. I shouldn't really be so damning about Henry's meanness, as I never paid Judy for the ring!

Henry was a regular visitor to Gus's elevenses along with the rest of the BBC outside broadcast team. He usually arrived with a terrible hangover and would respond to my asking if he wanted a drink with an initial shake of the head, followed not many seconds later with, "Well, perhaps a small pink."

His morning shakes, after the previous night's liquid intake meant that he needed a well-establish routine to get the first drink to his mouth successfully. His right arm would be wrapped around the tent pole holding him erect and a large gin and angostura would be placed in his right hand. He would then take a highly coloured kerchief out of his breast pocket with his left hand and tuck it into his right hand next to the glass. After this he gently pulled the other end of the cloth around the back of his neck, lifting his right hand and so bringing the glass smoothly up to his lips with nary a drop spilt. He insisted that it was the only way to get him started in the morning. After a couple of these he was able to join the others in a glass or two of Gus's Champagne.

He did, in fact, buy some cases of Champagne from us, which I would deliver either to the two Windmills, called Jack and Jill, at Hassocks, where he lived on top of a hill on the South Downs or to the Bath Club in Mayfair. He always had cases of Bollinger or Lanson Imperial Pints, sadly no longer made, and when I asked him why, he said, "They are a useful size, as it means that one can, should one be so inclined, allow the lady wife to have a glass." The first time I delivered some bubbly down to Hassocks, Henry fooled me in the same way he did every visitor. When I asked why only one windmill was turning he replied in his droll voice, "Well there's only enough wind up here for one." I nodded dutifully and only on the way back did I realise that I'd been had.

One of my most treasured memories of Henry was at Dalmahoy Golf Club near Edinburgh in the early sixties. We were at the Senior Service Festival of Golf and two days of non-stop torrential rain had brought play to a halt and the whole day had been cancelled. We were sitting in our kitchen surrounded by unused food and with rain starting to seep though the canvas. Henry came across from the Clubhouse at eleven looking for Gus, who quite wisely had decided to visit some friendly distillers in Edinburgh. I told Henry that Gus was not there but offered him a drink. He accepted with alacrity and proceeded to regale us with an endless fund of stories for the next four hours, lubricated by firstly the pink gin, then the Champagne, and finally, some malt whisky. We heard about his early years playing golf for his university, the time when 'the yips' stopped him playing golf any more and the realisation that his wonderful gift of words had a ready market that was being created with the televising of his sport, which took him profitably, and amusingly, all round the world.

He invariably used golfing metaphors to describe almost anything, as on the occasion when he was talking about the life-long battle he had with the Inland Revenue, "You know you are never going to win the match but you might, if you cheat a little, halve the odd short hole." Wittily though he might phrase it, the tussle was very real and he swore that he would rather die in Brighton Jail than give in to what he considered the totally unreasonable behaviour of the bureaucrats.

One of my final memories of Henry is in Gus's tent in 1978, when we were having a rather unusual 'Elevenses', as Gus and I were presenting magnums of a rather exceptional Champagne and specially engraved glasses to David Foster, the hard drinking chairman of Colgate-Palmolive who were sponsoring the event, and to Mark McCormack, whose company, IMG, were running it, to celebrate our one hundred and seventy-fifth golf tournament. David came in with Arnold Palmer who was going to be his playing partner in the pro-am that morning and Henry was already there with Mark and Gus. Henry responded to Arnold Palmer's cheery greeting, "How're you going Henry?" with a quiet, " Oh, just chipping onto the eighteenth green."

Henry and Gus putted out within weeks of each other a few months later.

5

UNINVITED GUESTS

Towards the end of my first season, Gus and I started to talk (although, to be truthful, Gus did all the talking) about how I was to be trained during the winter, when our away jobs were over and all the rest of his staff went largely into maintenance mode. He was a great believer in the practical approach rather than the academic and he certainly knew enough people to find training opportunities for me.

I definitely wanted to learn about wine and, since Gus's old friends at Booths Gin also, for some reason, owned a major wholesale wine merchants called Samuel Marts & Blundells, towards the end of October I went on the Metropolitan line train from Shepherds Bush to Farringdon Station and to the Clerkenwell cellars of this old wine merchants to meet the director, Jack Ford.

At that time London wine merchants still imported wine from all over the world in cask and bottled it in their own cellars. Samuel Marts were no different and so I had three months' training in the intricacies of storing, bottling and binning wine, including all aspects of the administration, Customs clearance, storage in bond, shipping and import documentation, as well as, most importantly, the tasting of wine. It was a fascinating time and was a marvellous basis of wine knowledge for the rest of my career.

Jack, who had been in the business all his life and obviously enjoyed passing on his knowledge, also told me a great deal about the tradition-bound way the wine trade worked in London (which was about to change dramatically). He talked about the concerns he and many of his long–established friends in the trade had about the fact that some less-reputable merchants were bringing in inferior wines from North Africa and stretching their French wines with them but, of course, maintaining the grand French names.

He was indeed quite right to be worried as within a dozen years this practice had become so widespread that the bottling of wines in London stopped altogether, largely under pressure from the French producers, and all wines were in future to be shipped over in bottle from the country of origin to preserve their good names, especially the French ones.

After that, of course, the stretching was done by the French in France (as always with the French, it was a question of who made the extra profit, not the wrongdoing itself, which merely caused a Gallic shrug). This was subsequently brought to everyone's attention by some very high-profile scandals in Bordeaux that stretched, if you'll pardon the expression, right up into the government.

After my time at Samuel Marts, Gus decided that my bar and restaurant service skills needed some further polishing up and so inevitably I ended up in one of his favourite haunts, the Belfry Club. This was in an old nonconformist church building in Belgravia, which had been converted into a very up-market dining club with a bar, and a restaurant in the well of the church. It also had tables on the wooden galleries at first and second floor levels. On the upper floors they had a few private dining rooms of differing sizes including the one in the tower of the building, the 'Belfry' itself, which seated about thirty and was where Prince Phillip had his bachelor party in 1947.

I worked first of all for a few weeks with Jack the bartender, a George Formby look-alike, but without the ukulele, who taught me all about mixing cocktails and remembering who drank what without being asked. He also speeded up everything I did, as his customers did not expect to wait.

The most interesting lesson I took away from those weeks was at the start of one morning when I was on my own in the bar, polishing glasses, and a member walked in. I knew him and served his Gordons and tonic, ice, no lemon, without prompting, and continued polishing while he talked to me. We had a very interesting talk for several minutes with no one else in the bar. When his guest arrived, he stopped talking mid-sentence, and instantly I did not exist. It was a real physical shock, although it shouldn't have been, but it taught me the divide between the server and the served.

Following the bar experience, I was to go into the restaurant to be a commis, working for the best waiter, Mario, at one of whose tables Gus usually sat. Mario's real name was also Jack – it seems my learning was all to be at the hands of Jacks-of-all-trades – and he came from Bethnal Green not Italy. He dropped his very pronounced Italian accent as soon as the customers left. I thought I knew what being a commis was from my time with Connie during the golf events, but this was something else.

Commis chefs in the kitchen and commis waiters in the front of house are the lowest form of restaurant life, worse even than the pot-washers; they at least do not exist as far as the senior staff are concerned. Commis are used quite literally as 'whipping boys' for the expulsion of all the venom that waiters and chefs built up during their time as commis, a sort of self-perpetuating miserarchy.

A commis waiter is the lad (or lass) who runs for the waiter and does all the jobs that the waiter doesn't like. He is occasionally allowed to serve the customers with simple things, but never anything that might get him noticed by the customer and in line for a tip. Mario had me dashing about all over the place and any shortfall in what he expected was rewarded with a clip on the ear or even, if no one was watching, a kick up the backside. It was quite astonishing. I was furious and yet this was the way it worked, like it or not.

My main task was to run up and down the spiral staircases to the kitchen to collect dishes for Mario's customers, and one occasion as I used the down-staircase I was passed by several commis going up at a hell of a pace, none of them carrying anything. I didn't register what was going on until I got to the kitchen, to discover that one of the sous chefs, about six foot six tall, built like a mountain, had suddenly gone totally berserk and was attacking anyone in a white jacket with anything he could get his hands on. Since the rapid departure of the other commis, I was the only white jacket in his view and as a chef, the things he could get his hands on included knives, cleavers and various pots, all of which he hurled at me, cowering at the foot of the stairs.

Luckily none of the items thrown hit me sharp end first but it was only time before one would. I was starting to think that perhaps I should have gone to university after all, where a cricket ball was likely to be the worst thing thrown at me, when two other chefs, much smaller than him, jumped up and, one on each shoulder, pulled him down to the ground. Then the head chef slapped him firmly a couple of times round the face and they all got up and carried on as if absolutely nothing had happened.

A very frazzled commis went back upstairs to be given a severe bollocking by Mario for taking so long and on explaining what had happened; he said, "You should always expect that, as all chefs, whatever sex, suffer from permanent PMT." After forty plus years the only alteration I would make to that summary is to change "all" to "most" (let the many chefs I have known argue among themselves about who has been excluded). I wondered what on earth Mario would have made of the fact that a chef, Anton Mosimann, now owns the Belfry.

Each morning during the time I was training at The Belfry I had driving lessons and the day after I returned to work at Shepherds Bush, I took my driving test. I passed and to celebrate Gus suggested taking me out to lunch. I asked if we could go to the Belfry. I wanted that bastard Mario to call me sir.

As we got into Gus's Humber he suggested that I drive, which was a rather tricky change from the BSM mini in which I had trained, as was the West End from suburban Ealing where I had my lessons. I managed to get to Belgravia and the Belfry Club without mishap and as I drew up outside the club, the linkman smiled as he took the car away, having got to know me quite well during my time working there.

We were shown straight into the restaurant and to Mario's table and as he held our seats for us while we sat down he said, "Good morning, Mr Payne - Mr Payne," nodding to both of us in turn without the slightest inflection to indicate that the second Mr Payne had been, only twenty four hours earlier, the object of his boot for not being quick enough in bringing a dish from the kitchen. I was now an esteemed customer. A bastard he may have been, but a very professional one.

So now it was time for me to go back to work at Shepherds Bush as a driver, which gave me a lot more scope. I arrived back to discover a new arrival in the team at Pennard Road, Pepe. Apparently Pepe, who not surprisingly came from Spain, was given a job as a driver by Gus as a favour to a Spanish friend of his, the director of a Sherry House. Pepe's arrival here had been very rapidly arranged and, judging by the interests he demonstrated during his stay, he was almost certainly trying to stay one step ahead of the Castilian father of a beautiful señorita; he certainly didn't come to improve his English, which was diabolical.

Pepe, who was small, wiry and with very handsome Latin looks, worked as a driver and porter and most of the time was trying to get under the clothing of any and all of our female staff, whatever age or shape. He was hard working and amusing, but only in mime, as he showed no interest whatsoever in learning to speak English; with hindsight he gave every impression of being the model for Manuel in Fawlty Towers a decade later.

About two months after his arrival we were working at the Royal Show and it was decided that Gus was going to have a dinner party in the ICI pavilion, where we were doing the catering, after the show had closed to the public. A lovely part of all the events where we were catering was the activity on site after the public leaves when the facilities are all ours – it is an enchanting time. This dinner party was to be for a dozen of Gus's old friends from the world of the agricultural shows, a circus-like troop who went from show to show during the summer and included the bosses of the agricultural departments of ICI, Shell-Mex & BP Oils and Shell Chemical, all of whom Gus had catered for over many years.

It was planned for me to look after the party with Connie, when Pepe let us know that he could wait at table, and had done so in Spain. He did this not by telling us but by getting dressed in his waiter's garb, which we had no idea he carried with him. We agreed that he could help me and so give Connie a night out with the many friends she had at the show.

While we were laying-up for the dinner, Pepe suddenly asked me, "How you English say like the French do, 'Bon Appetit'?" Seeing an opportunity for some fun I replied, "In English we say 'I hope it bloody-well chokes you'." I then, with a perfectly straight face, rehearsed him in this until he had mastered it.

The first course of the dinner passed without incident and I then served the meat course, leaving Pepe to finish off by serving the vegetables. As he finished serving my father last, Gus looked up over his half-glasses and said, "Thank you Pepe", and was totally non-plussed by the unexpected and courteously offered comment by Pepe, "Thank you sir, I ope eet blodywell chocks you." The almost apoplectic effect on Gus and his guests was wonderful – they had no idea what to think until Gus caught sight of me rolling about with laughter just inside the kitchen door. Needless to say, Pepe was none too happy with me after that. He stayed with us during the whole of the summer until an accident forced us to rid ourselves of him.

We used to cater regularly for lunches at an experimental farm Shell Chemicals operated deep in the Buckinghamshire countryside, where they tried out their fertilisers and other products and showed them off to small groups of farmers to whom we served lunch in the farm buildings. A couple of times a year they held Open Days when a marquee would be erected for a week and a hundred and fifty to two hundred farmers and farm managers were invited to a day out including a slap-up lunch.

I was on site with Connie on the second day of one of these sessions, having set the tent up and served breakfast to the Shell Chemical staff and the farm workers. We were all ready for the lunch and waiting for the assistant chef, who was bringing the additional supplies, and the waiting staff, who were being driven down to the farm in the minibus by Pepe.

Pip Thow, the assistant chef, was in the van with the top-up food and driving along a narrow country lane down a steep hill when he was stopped by a queue of cars in front of him. After a couple of minute's inactivity, since he was in a hurry to get to us, he decided to walk down the hill to find out what the problem was. Thirty cars ahead he saw, to his horror, that the cause of the logjam was our minibus, which had turned over and crashed through a wall at the bottom of the hill. He ascertained that the staff, while damaged, were not seriously hurt and went into the pub whose garden wall Pepe had re-engineered.

He phoned the office at Shepherd Bush and eventually the message came though to the farm office and a runner rushed across the few hundred yards to the marquee where I was told of the problem – once again, wouldn't it have been nice to have had mobile phones – and Connie, Jimmy the chef and I discussed what we were going to do. Jimmy had enough food for today's lunch as Pip was bringing mainly the third day's supply, but we didn't have any waiting staff.

Connie set to as ever. She got the farm boys in and made them wash and scrub their hands, which she then examined. This involved them all standing in line facing her with their hands out front as she, as the WRAC sergeant she had been, examined their hands and nails both sides – a very amusing sight, this line of large, totally sheepish men being cowed by this small and extremely determined woman. When she was satisfied, she and I started a very rapid indoctrination of these big, galumphing fellows into the finer points of serving the guests who were, while this was happening, touring the farm listening to the explanations of crop rotations and so on and looking forward to their lunch, blithely unaware of the hiatus.

The astonishment of the guests at having their meal served by these lads was nothing compared with the moment when the replacement minibus arrived and a trail of waitresses wandered aimlessly into the lunch tent, cut, bruised and battered, the older ones looking seriously ghoulish with large streaks of blood through their white hair, the whole troop having the demeanour of Napoleon's Dragoons returning from Moscow.

It was at this point that Roger Clark, the one-eyed farm director, stood up to tell everyone what had happened and spontaneously everyone roundly applauded both the wounded tribe of waitresses and the newly-talented farm labourers. Roger Clark said afterwards that the episode gave him a useful diversion, as his guest of honour, next to him, had been deaf in one ear and, as Roger only had one eye, they had spent the entire lunchtime jockeying for position to carry on their conversation.

The waitresses told tales of Pepe's reckless driving at

all times and especially on the day of the crash. This was in spite of the fact that they liked him and not all of them were averse to his fumbling, youthful groping (I knew of this because I heard them chatter about it when I was driving them to and from jobs). Needless to say Pepe had to go – and we never heard whether he went back to face the irate Spanish father-of-the-would-be-bride of our imagination.

I spent the summer pretty much following the routine of my first year, primarily looking after Gus's guests on the golf tournaments over the lunch periods but gradually given more responsible jobs as well, such as helping Big Alan sort out, allocate and instruct the temporary staff at the beginning of each event.

This more supervisory role posed a new problem. I had a very young face - not surprising since I was very young, but it didn't sit well with telling these worldly-wise old birds what to do. I also suffered from a very fast growing beard, which I had to shave twice a day causing considerable discomfort.

I solved the two problems by growing a beard. This has been a permanent addition ever since, having been taken off only twice in the subsequent forty years. I needn't have worried about looking too young as within a couple of years the hair on my head started to disappear. Coupled with the beard, this gave the appearance that gravity had made the whole thing just slip round my head.

The first occasion I removed the beard was once when I decided that the seven- year-old beard should go as we were about to go on holiday and I could therefore re-grow it if the result was not good. I went to the barbers in Shepherds Bush Market and returned to the office clean-shaven. For a good half hour no one noticed. There were then some squeals as my colleagues slowly came to realise how different

I looked. I then went home and my daughter, Emma, one year old at the time, burst into tears as she recognised my voice but not my face. I grew it back almost immediately on the holiday.

The only other time was again at the beginning of a holiday, in Portugal, when in the presence of Judy and our teenage children I removed the beard, only to have my fourteen-year-old son Richard comment on the finished article, "Good God dad, you look like a baked bean." And he was absolutely right, so back it came, never to be removed again.

Richard, endowed with his grandfather's charm, was always able to deflect criticism with a witty, usually self-deprecating remark. The most memorable was this same year when, Judy and Emma being away, I was looking after him. He was getting over flu and was well enough to get up but not to go back to school. Thoroughly fed-up with himself, watching television, he said in a pathetic voice, "Dad, make me a hot chocolate". Being a caring sort of father I responded, "Make your own bloody chocolate", only to be told, "That's the trouble with you dad, you were never brought up as a spoilt child." Needless to say he had his chocolate made for him.

Back in the early sixties, the great thing about looking after Gus's hideaway was not just learning my trade but meeting the fascinating array of guests, who, by the nature of Gus and his business were a mixed lot, but nearly all involved in running, broadcasting or sponsoring sport, primarily golf.

The golf industry guests included a regular team headed by the imperious Commander Roe together with Dick Penfold, whose golf balls rivalled those of Dunlop, Buzzer Haddingham, boss of Slazenger and subsequently the Chairman of the All-England Lawn Tennis Club at Wimbledon and the Dunlop Golf supremo, Eric Hays.

Commander Charles Roe was over eighty when I first met him and he was still the Secretary of the Professional Golfers Association. He ran this organisation on his own with just the assistance of his secretary, Miss Cockburn, in a tiny Dickensian office in Ethelburga House on Bishopsgate in the City of London, now no longer there. He managed all the tournaments from April to October and also looked after the affairs of the Club Professionals.

In order to continue the tax-efficient concept that the sponsors' entertainment was part of their sponsorship, and to ensure that he was never left holding the bill, he would have each sponsor pay into the PGA's account, before the event, all the money needed to run the entire tournament, including the catering, tentage, accommodation and all and sundry. We, like all the other suppliers, would submit our accounts to the PGA and receive a cheque the next day.

Knowing this, Gus would, with my help, work on the account on the Sunday after a tournament and very often on the Monday I would be sent on the Central Line to Liverpool Street with the bill and the box of lunch, tea and drink tickets, which were our currency for the fare we served to sponsors' guests.

I would go up to the fifth floor in the Victorian lift and tremulously knock on the door. Welcomed in by the Commander's booming voice, I was initially quite terrified of him with his ramrod back, massive bald dome and piercing blue eyes. He used to be a submarine commander and I cannot imagine any sailor under his command getting away with anything, nor even trying.

It was very difficult to remind oneself, in the early sixties, when seeing him still in full use of all his faculties and in absolute control of professional golf, that he had retired from the Royal Navy in nineteen thirty three. He was always very pleasant to me and having taken the invoice, would say, "I don't think I need to count the tickets, but I'm sure you would like to save me the expense of a stamp by kindly taking the cheque back to your father." He would then hand over the invoice to Miss Cockburn who produced the very welcome cheque that I banked when I got back to the Bush and which kept us going for quite a while.

At the eightieth birthday party that Gus laid on for him in our cellars under the warehouse in Pennard Road, Charles Roe informed all the assembled company that he had worked out that since he had little income from the PGA he would have to tap into his savings and these would run out by the time he was eighty-five, and so they had better arrange to have him shot at that time. Remembering this five years later, Gus organised a lunch for his eighty-fifth birthday, again in the cellars. After lunch Gus took him out into the area below the pavement where a post had been set up to which he was to be tied and all the usual collection of his friends who were guests were each given a toy rifle. The "Shooting Party" became an annual event right up until his death at the age of ninety-four. I particularly remember his ninetieth, which was done in the style of the Gay Nineties in Paris and I dressed up in frock coat and top hat and down on my knees as Henri de Toulouse-Lautrec, rapping for attention with my silver-topped ebony cane and ordering them to lunch. It was going to be twenty-five years later that I would be similarly togged up, although not on my knees, selling our wares in Chicago.

The other two great teams who joined Gus at the tournaments for Champagne Elevenses, and very often stayed on for lunch, were the BBC Outside Broadcast producers, directors and commentators and a very convivial group of tournament golf professionals. Most of the BBC team were known as names on the credits of programmes such as Sportsview and Grandstand, rather than as famous personalities, but there were some powerful and lively characters such as Peter Dimmock, Ray Lakeland, Slim Wilkinson and Ginger Cowgill. They were initially introduced to Gus's morning ritual by Henry Longhurst, of course, but once aware, needed no further prompting and, particularly in the early days when television coverage was only on the first and last few holes, there was a lot of the day to fill in and our problem was sometimes getting the buggers out.

On one famous occasion Gus went out of the very full tent and said in an imaginary conversation, "Hello Peter, how are you and how's Polly?" The thirsty BBC crew inside the tent, assuming Gus had met Peter Dimmock, very much their boss, and was about to bring him in, all dashed out through the kitchen so he shouldn't catch them. A great ploy, but you can only use it once.

The crew of golfing regulars were led by Peter Alliss, whom I had first met at the tennis at Bournemouth when he was the pro at Ferndown Golf Club. Peter's great pal on the golf tour was Dave Thomas, with whom he subsequently designed many golf courses including the Belfry Club, not the one I trained at but the one at Sutton Coldfield, scene of many Ryder Cup Matches. Dave was a massive chap with a most prodigious appetite which we had great difficulty satisfying. Another regular was Christy O'Connor who was always in favour of drinking some 'red biddy' with his lunch regardless of whether he still had to play his round. Christy came in one day sporting a beautiful black-eye, a present from Harry Weetman, the result of a fracas in the clubhouse the previous evening which had also seen Harry's wife, Freda, laying out the captain of the club with her handbag.

Max Faulkner, famous for nearly thirty years as the last British winner of the Open, regularly came in with his extraordinary gruff voice demanding with his lunch, "Pot of tea, please, guvnor, very strong." I knew that he always wanted a 'road-builder's tea' strong and sweet enough to stand the spoon up in it.

Gus also provided a welcome for several visitors from overseas when they were here for tournaments: Henry Cotton, Peter Thompson, Bobby Locke and latterly, Tom Watson, became firm friends.

The regulars would also very often bring in their friends and as caterers it was through these introductions that several interesting, and often quite profitable, parties came our way. Peter Alliss started the most notable and hilarious of these when he brought in a man called Johnny Speight, the writer of the outrageous and very popular television programme 'Till Death Us Do Part'; in other words the man who created the wonderfully dreadful Alf Garnett.

This happened at the Piccadilly Medal Tournament, which was being staged at Princes Golf Club at Sandwich. It was a particularly memorable event because Tony Jacklin was playing there for the first time since winning the Open Championship at Lytham the previous month. His wife Vivienne was with him and was expecting their first child at any time. Judy was in almost exactly the same mode; in fact the Jacklins' son Bradley and our daughter Emma arrived within days of each other less than two weeks after the tournament. Judy and Vivienne spent the week waddling around together, bemoaning the exceptionally hot summer and enjoying each other's company. Vivienne was delighted to have someone to talk to who was not directly involved in golf. Her husband had, by being the first British winner of the Open in thirty years, suddenly become everyone's property. In a fortnight her life had been turned upside down, she was confused and disorientated but they had also become very wealthy.

They were in Gus's tent on the second morning of the event when Peter came in with Johnny Speight, who was about to change the week for us. Peter is, as anyone who has listened to his subsequent television commentaries is aware, a very funny man and can 'hold court' in any gathering and have everyone in stitches. However, I have noticed over the many years that if there is someone present who Peter thinks is funnier, he is unaccustomedly quiet and very sensibly defers to them. He has a hilarious golf-pro friend called Hugh Lewis with whom Peter always takes a back seat and listens. It was certainly so with Johnny Speight.

Johnny was a natural cockney comic, quick-fire in spite of a rather engaging stutter. He kept the entire tent enthralled all day so that no one seemed to want to go and watch the golf. He told us of his totally uncontrollable children sitting in the back seat of his Rolls Royce Corniche and touching the button on the side of his seat moving him backwards and forward while he was driving. He said, only half-jokingly, that he had become so important that each time the Director-General of the BBC saw him he went down on one knee and kissed the ring on Johnny's finger. Johnny enormously enjoyed my interjection, "That's because the stone in it is a garnet." He chuckled endlessly about that and I knew it was going in the repertoire.

The whole time we were worried that either Vivienne or Judy was laughing so much they were liable to give birth in the tent. Given that treatment, it's no wonder that our delightful daughter Emma came out smiling and has been 'crackling' with energy ever since – not all of it welcome to her parents!

We then joined Johnny as his guest for dinner in the Bell Hotel in Sandwich where he and numerous people from the tournament were staying and, besides keeping us amused and the serving staff in a state of total confusion, he also told Gus that he wanted him to cater for a party for him at his home in Northwood, at the end of the month. He outlined what he wanted in very amusing style and when Gus told him how much it might cost his only specific instruction was, "N-n-o, n-n-no G-G-Gus, I w-w-want to spend m-m-m-much m-m-m-more than that!"

After dinner we went into the lounge and were joined by several others for coffee and liqueurs. A tremulous young man arrived to take our orders and it was quite obvious that his day had suddenly become very difficult with this lively, talkative group all ordering a bewildering variety of drinks. Among the items he made a slow and deliberate note of in his little book was Judy's choice, crème de menthe frappé. He then staggered off towards his bar and a good fifteen minutes later he came back, seriously looking at his book and asked in a never to be forgotten phrase, "What's crème de menthe crappé?" Following the inevitably uproarious reaction to that, it took quite some time to get all the drinks, but after we did, we agreed the date for Johnny's party with him and Gus said we would visit the house and then send him a detailed quote, all of which duly happened.

When we went to the house, we discovered from talking to the housekeeper that, not surprisingly, the neighbours had not taken to him. Northwood at that time was very snooty and Johnny delighted in puncturing pomposity, so they were not going to welcome him. Johnny, of course, with his Corniche in the drive, which none of his neighbours could afford, played the game for all his worth. We were told that Johnny's kids, who were neighbours' kids from hell, had kicked their ball into the next-door garden. Johnny instructed his city solicitor to send a formal letter to the neighbour's solicitor, asking if they could have their ball back.

So correspondence went from us to him and from his secretary back to us and we were told that nobody would be at the house when we arrived but were told where the key would be. We arrived at lunchtime and spent the afternoon setting up the party for a hundred or so guests who would arrive from eight o'clock onwards and eat from nine-thirty. When Johnny arrived home about seven, he said, " Wwwwhat the fffffuck are you lot doing 'ere!" We, naturally, thought he was joking but he wasn't. He had totally forgotten about the party. What's more he had forgotten to invite any guests! There followed the craziest evening with him phoning friends like Jimmy Tarbuck, who at the time was doing a summer season in Margate, who promised to come up with the whole cast after his evening performance, which he duly did. However in the days before the M25 the journey from the Kent coast to the stockbroker-belt of North-West London took them until way past midnight.

While we sat around, amidst the vast amount of food we had brought in order to justify the large sum of money Johnny had insisted he wanted to pay us, he went down to the local pub and got everyone, including the landlord, to join in the party.

So it was that, somewhat later than planned, the guests started to arrive and Johnny gave Connie, whom he adored from the time he first saw her in Gus's tent, her instructions for the evening, "Just b-b-bring me a large w-w-w-w-whisky every ten minutes, w-w-w-whether I ask for it or not, ok?" Eventually after a great many of the ten-minute interval ministrations by Connie, Johnny fell asleep in his chair. The uninvited guests continued to enjoy themselves till morning.

6

Step From The Shadow

When one has a very dominating and charismatic character as a father, life is never easy. When that person is also one's boss it becomes considerably more difficult.

I felt a great deal of sympathy for Kevin Maxwell when he was tried for being involved in the sins of his father, the egregious tycoon Robert. I thought of writing to the court to say, "No matter how old he is, there is no way he could have refused to do what his father told him." – no defence in law, I'm sure, but true all the same. This belief came, not just from having started my own business life under the influence of a powerful 'life force', but also from having seen Robert Maxwell at work and the domination he exerted over his family.

In the early eighties, we undertook a party for Robert Maxwell to celebrate his company, Pergamon Press, taking over the British Printing Corporation. The party took place at Headington Hall, the mansion he rented from Oxford City Council, and as such the grandest council house in the country.

After we had set the whole thing up the day before the party itself, we were summoned to the great man's study to be confronted by the tycoon himself, massively seated behind a very impressive desk. What made the scene even more dramatic was seeing his entire family standing in a semi-circle behind him, like a Victorian melodrama.

He barked questions at us in his resonant too-perfect-English voice, and subsequent not-to-be-denied instructions. This was fine, we were caterers brought in to do a job and being well paid for it. Of course we were used to being told clearly and forcefully what was expected of us. There were three of us 'in before the beak' as it were: Alan Cox, who was the director in charge of the function, a tall lean young lad called Bob Russell who had only been with us for a year or so at the time, and myself. I was not directly involved with the event but, as the boss, my presence was demanded.

Alan Cox, the same height as Maxwell, stood ramrod-straight in his black-jacket-striped-trousers-caterer's-uniform, his very large ears glowing red was the only sign of nervousness. The staff used to call him 'Dumbo' behind his back, but very quietly as, not surprisingly, his hearing was very good - and he was a big chap. He was normally extremely good at dealing with clients as, after thirty years of organising such parties, nothing fazed him. This however was of a different order altogether.

Maxwell had just flown in so he had a plethora of very detailed questions about the staff, service, where the food and wine had been purchased, and every time we answered he came back with supplementary questions. I noticed that members of the family were taking copious notes. Very specific questions about the wine were prefaced with a demand to know who was going to be in charge of wine service. Alan Cox told him that that was Bob Russell's role; so Bob then received the deluge of question after question, demand after demand about the wines and their purchase, preparation and service. A couple of times Alan or I tried to answer on Bob's behalf, but were firmly told to shut up. Bob, never a ruddy-hued young man, went a very pale shade of grey under the onslaught, but coped admirably.

Besides the detailed questioning, what was decidedly different about this occasion was the way that he barked out orders to the various family members and was exceedingly harsh and condemnatory in the comments he made to them, in front of us. It was seriously embarrassing for us, but quite apparently not the least bit of a surprise to them. They did precisely what they were told, without comment.

Taking a peek around the house was of course, irresistible, and we could see signs of the tycoon's megalomania, most notably in the meeting room, which had a vast table surrounded by twenty, seemingly identical, large white leather armchairs. However each of them was in a fixed position facing Maxwell's chair at the top of the table. What's more, as a visitor sat down the cushion gave way so that he sank into the depths of the armchair, whereas Maxwell's chair was larger than the others and the seat cushion held firm leaving him, already a formidable presence, raised up, and even more intimidating.

The party was one of the most unpleasant functions we ever organised. A feeling of fear pervaded the entire event. The British Printing management were less than delighted at becoming part of Maxwell's empire, little knowing that one future benefit would be having their pensions raided.

Although Alan Cox had organised some previous functions for Maxwell, I had never met the great man before. As it turned out, we would not do any more catering for him. We discovered the reason for the questioning and the family taking notes was that it was his intention to get his own staff to do any further parties. He was taking this opportunity to pick our brains.

The only other occasion I encountered him was at the Bob Hope Classic golf tournament at Moor Park during the autumn of the following year. It was quite a cold day and Robert Maxwell was wearing a massive camelhair coat against the chill as he walked around the golf club. He went up to a man selling the London Evening Standard and, with his vast podgy hand, reached out to take a paper. When the paper man put out his hand for payment, Maxwell offered him a fifty-pound note. Whereupon the paper man said, "I can't change that!" Maxwell replied, "Suit yourself ", took the note back, and went off jingling change in his pocket. The millionaire, who that year had purchased the Mirror Group, not only cheated the paper man out of twenty pence, he made sure that he knew he had been cheated. At that moment I was very pleased that we would not be doing any more functions for him.

My father, while dominating and charismatic, was nothing like the ogre Robert Maxwell was. Yes, Gus did bully, but he did it by charm rather than threats. Back in the nineteen sixties, the biggest problem for me was trying to get out from underneath this very powerful personality and establish myself. I wanted to be Alan Mark I rather than Gus Mark II.

One of the first opportunities I had to be away from his effect was, ironically, brought about by a typical piece of Gus magic. In the spring of 1964 he had taken me out to lunch at a small bistro in the basement of a house in Dean Street, Soho. Apart from the grand clubs that he frequented, Le Caveau was one of Gus's favourites, largely because of the

'Little Alan' at 16

Gus Guffaw

Alice Payne's Coffee Stall 1908

Alice Payne's Café Shepherds Bush Market

Gus and Lila

Gus the Waiter

First Event Home Guard Dinner 1946

Gus and Dai Rees

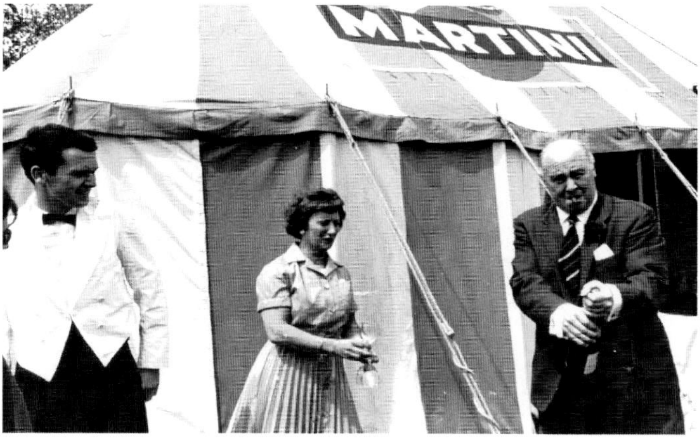

Elevenses, Gus, Connie and Little Alan

Commander Roe's 80th Birthday Pennard Road Cellars

Henry Longhurst's 'Tented Village'

'Little Alan' at 33

April Fool

Gus at Alan and Judy's Wedding

Paynes' Pavilions at Henley Royal Regatta

John Coomb the City's Head Waiter presented with Catey Aawrd

authentic French country food, and the jolly, Beaujolais-drinking Gallic patron who fell upon Gus each time they met as if they were old comrades in arms. A few years later this patron had, as usual, gone away for the weekend to his country retreat with his English wife, who helped him to run the restaurant. Gus, and all the other regulars, were completely bowled over by the subsequent news that on this occasion he murdered his wife and then returned to run the restaurant for a couple of days before killing himself. Not just a jolly Frenchman, but a jolly strange Frenchman.

At our lunch, Gus inevitably struck up a very amusing conversation with an elegant Saville Row suited man on the next table. He turned out to be Tony Reuss who, with his brother Ben, was the importer of, among other things, Pol Roger Champagne. Gus told him that I had taken an interest in the wine side of the business and before we knew what was happening, and before we had finished lunch, Tony had invited me to be the Pol Roger candidate for the Champagne Academy that autumn. I knew nothing about this organisation but discovered that the twelve top Champagne houses, the so-called 'Grands Marques', each nominated a candidate, through their agent in London, to participate in a course studying Champagne. This entailed a week in Epernay and a week in Reims, staying with the Champagne families themselves, with a weekend in Paris observing Champagne being drunk at the Lido Night Club. I gathered that it also involved the students drinking a certain amount of the wretched stuff.

At the end of the fortnight, if an exam is passed (that is, if your pen hits the paper), you become a member of the Academy and your subsequent role in life is to promote the good name of Champagne. It was, needless to say, an arduous couple of weeks but someone had to do it, so it might as well be me.

The rotating presidency of the Academy was held that year by the House of Louis Roederer and so their agent in London had the responsibility of arranging everything for the trip. Thus it was that a dozen young men from various parts of the wine trade made their way on a Friday evening to the City of London. Going up the marble steps of a forbidding Victorian building within sight of the Monument to the Great Fire, they went past the brass plaque from which countless cockney cleaners had almost polished out the name, 'Rutherford, Osborne and Perkins - Wine Shippers'.

As a typical caterer, I was the first to arrive. I was greeted at the top of the highly polished curving staircase by David Rutherford, latest generation of this family wine shipper, and destined to be the last independent owner. Ruddy faced, with premature white hair fluffed out at the side of his balding head, and a nasal twang to his upper crust tones, he had all the appearances of an archetype of the London wine trade and, as it turned out, was an extremely talented man of wide and amusing interests.

I discovered that David's father, Colonel Jack Rutherford, was one of the beacons of the wine trade. He was at that time on the panel of the Queen's wine advisors, so David and I had a lot in common over the years, having both followed powerful and well-respected fathers in business. This common purpose was added to when Rutherford Osborne, a few years later, was taken over by Martini & Rossi and David joined their board. Martini was one of Paynes Caterers' biggest clients as we did the catering for most of their many sponsored events, ranging from golf to archery and motor racing to fencing.

On this first meeting, as we waited for the arrival of

my fellow students, he took me through to the large corner office with bay windows overlooking the Monument, and introduced me to three more of the agents for the Champagne Houses. The party who were to visit Champagne included the scions of a couple of the large regional breweries that still played an important role in the wine trade. There were also middle managers from some of the more aggressive, publicly quoted, wine companies, as well as representatives of the next (and in some cases the last) generation of small, independent wine merchants, together with the nineteen-year-old son of an outdoor caterer from Shepherds Bush.

As each new candidate arrived, David welcomed him by name and recited the details of all the assembled company without any hesitation or a single mistake. When we had all gathered, with the first of very many glasses of Champagne in our hands, he gave us details of the trip and what was expected of us as representatives of the English wine trade. He also mentioned that the previous contingent had been the first one consisting entirely of the Irish Wine Trade, so we had a lot to live up to!

The following morning we were flown by private plane to Reims airport, where we were met by André Rouzaud, the President of the Academy, and the elegant owner of Champagne Roederer. He was accompanied by half a dozen other proprietors. It was a very heavyweight contingent, which identified how important they considered our visit. This was to set the tone for the entire fortnight. After a glass or two at the tiny airport we were driven to the vast Art Deco building in Epernay that was the headquarters of Moët & Chandon, the biggest of the Champagne houses. I had never seen a large French office before and the sight of clinically white-coated staff busying themselves, click-clacking around the marble-floored, glass-encased offices reminded me of Jacques Tati's wonderful film, Mon Oncle.

We then were allocated our 'digs' for the stay in Epernay. A couple were with Pol Roger, next door to Moët on the Avenue de Champagne; a couple more next door again at Perrier-Jouet; two more at Bollinger in the lovely village of Ay. A further three stayed at the magnificent Chateau de Saran, the seventeenth century hunting lodge that Moët had turned into a wonderful entertainment facility for visitors including, over the years, monarchs, heads of state, film stars and other celebrities.

The last three of us were accommodated in the best hotel in Epernay, Des Berceaux. Though we three initially thought ourselves short-changed, we actually had the best of both worlds as we joined up with the others for dinner at the various Houses during the week but were then free to take advantage of the surprisingly lively night-life in the town.

Although the whole Champagne trip was a cooperative effort by all the Houses, they all tried to outdo each other in the entertainment they gave us. We had a glorious time and the first week is somewhat of a haze, but two memories stand out.

The first was dinner at the private house of Madame Lily Bollinger in Ay. Hers was a modest home, without the vast number of servants enjoyed by several of the larger houses. She had for this occasion employed a local caterer to look after us and when we arrived and were given our inevitable glass of champagne, it was obvious that things were far from tranquil.

Madame Bollinger, a formidable personage, probably in her seventies at the time, and in the grand tradition of successful Champagne widows, was, while charmingly attentive to our needs, more than somewhat distracted and we couldn't understand why this should be. Then the news was whispered round the room. It appeared that the caterer hadn't arrived with the decorated salmon for our supper, which was enough to make any hostess agitated. The salmon, accompanied by a red-faced caterer, duly arrived, and I have to say the fish was well worth the wait. Given the connection we were to have with the House of Bollinger in future years at the Open Championship and elsewhere, it was uncommonly decent of them to make me, an English caterer, feel so at home by staging such an episode.

Although I'm sure she didn't find it amusing at the time, she laughed ruefully about it when, several years later, we talked at a party I catered for her outside Bollinger's London office in the charming Georgian courtyard, Pickering Place, in St James's. The reason for the party was suitably eccentric, to launch a rose created by Harry Wheatcroft and named Lily Bollinger.

The other outstanding memory of the initial week in Epernay was my first view of the Chateau de Saran. After a morning of being taken round the many kilometres of cellars under the Moët offices, we were then taken another few kilometres out of town to this fairytale chateau, an old hunting lodge. In their wisdom Moët had refurbished this building to what it would have been before the revolution or certainly at the time of Louis Napoleon. Opulence of a standard we Britons were certainly not used to. With gold-encrusted rococo decoration and elegant First Empire furniture, it was presided over by a priceless treasure of a British châtelaine. Over many years the Honourable Moira Campbell turned Saran into one of the great places of entertainment in the world. She was quite wonderful: imperious, elegant, and beautifully dressed, always in a totally appropriate style for the time of day and for the guests she was looking after. She could talk to anyone about anything and certainly made all of us feel very welcome in what could have been quite intimidating surroundings.

The white-gloved waiters, in their traditional charcoal grey champagne aprons, with the Moët logo on the breast, offered dish after dish of outstanding and vastly impressive food. This was the first time that I had experienced this style of service where the tray is proffered alongside your plate with spoon and fork available for you to help yourself. Known as 'butler-service' and traditional in country houses, it was no longer common in England, and little did I know it would nearly lead me into a disaster a quarter of a century later.

Each dish was refreshed and re-presented a few minutes later and Moira, who was served first, would always have a second tranche, firstly to make the guests feel comfortable about having more, but also to make sure that the chefs didn't have one of their creations returned untouched. We understood that each winter Moira retreated to health farms in the UK to repair the ravages to her liver and waistline. She then returned to the fray the following spring, and she continued this routine for more than twenty years.

I have to say that I greatly enjoyed the company of my peers in the trade, most of them considerably older than me. I also thoroughly relished meeting and staying with the proprietors of the Champagne Houses, Paul Krug, Victor Lanson, Prince Guy de Polignac of Pommery & Greno, all of whom were charming as befits the patrons of a product such as Champagne that is so redolent of entertainment. The wonderful thing was that they looked after us personally. And all this without Gus, it was marvellous.

Our last night in Epernay was a particularly long and onerous one. It involved a variety of locations, numerous bottles and, with no lectures the next morning, merely lunch at the golf club and a drive to Paris; we gave in to the temptation and ended up in a nightclub called Le Jockey.

I have no recollection of returning to the hotel but I must have done so as the three of us were collected from there the following morning with our cases for the trip to Paris. We were chauffeur-driven in the outrageously ostentatious silver Cadillac of the Academy President, André Rouzaud. What we had forgotten was that, although we were not having a lecture that morning, we were visiting the factory where the bottles were made. When one has a hangover induced by a week of overindulgence followed by a night without sleep, the last thing that one needs is to walk round a factory with furnaces and machines cranking out bottles. The effect of the heat and the noise would have made one think of Dante, if one had been able to think. The worst moment was when we were shown a bottle tested for its strength by being filled with water and having air pumped into it until it burst. The anticipation of watching that bottle and waiting for it to explode, in the fragile state we were in, was an exquisite torture worthy of Torquemada and his fellow Inquisitors.

From the glass-works we were taken in the Cadillac to the Golf Club de Reims where a sumptuous lunch was laid out for us. However, three of us were in no condition to face any lunch, let alone sumptuous, and went for a walk around the golf course. When we arrived back at the club, a silver Cadillac was parked by the chateau-style clubhouse and the driver was just opening the boot. I assumed that it was our car and that it would be rather nice to sit in it and go to sleep before we were driven off to Paris.

So I said to the rather elegant man, "Are you our driver to take us to Paris?"

The response became the story of the group thereafter,

"No, I am Count Robert-Jean de Voguë, owner of Moët & Chandon, and I'm here to play golf." He chuckled at our mistake saying, "I'm sure that André Rousseau would be delighted to have me taken for his chauffeur, but tell me are you not supposed to be at lunch in the club?" We explained why we weren't and again he laughed, "It's like hitting your thumb with a hammer – you know it's going to hurt but still you do it, n'est-ce pas?"

It would appear that the weekend ahead in Paris was going to involve a great deal of thumb hitting. And so it was, as our very effective pied pipers for this escapade were to be Francois and Jean-Baptiste Lanson, two of the younger generation of the Champagne family.

The highlight was the evening at the Lido, the extravagant nightspot that was the template for virtually every other around the world. We were ushered past the long line of guests waiting to be shown to their places and led to a long table right at the front and leading away from the stage area. I was put right by the stage, which I found quite disconcerting, but it was my twentieth birthday so I was given the honour. What I didn't know, but the Champenois certainly did, was that after dinner when the show was about to start, the stage lifted up four feet and my chair was on it. So there I was, four feet above everyone else, in a spotlight, while a team of waiters brought in a cake and the entire theatre sang 'Happy Birthday', embarrassing, but very enjoyably so. Luckily, or unluckily, the cake did not contain a 'Bluebell Girl'.

The following day, nursing yet another grievous hangover, we quietly made our way around Paris, which I had previously seen at the age of fourteen, on a school exchange between Latymer and the Lycée Chaptal, in very different circumstances. Lunch in a restaurant in the Bois de Boulogne was followed by a gentle train journey back to Reims in readiness for another week at the coalface.

However, before that we were, according to the itinerary, due to be given a meal in the Buffet de la Gare – the station buffet. Compared with the extravagance and elegance of all the meals we had been served so far our taste buds shuddered at the prospect of a railway buffet, which in England in the early sixties conjured up thoughts of the café in 'Brief Encounter'. We needn't have worried; the astounding display they put on for us was as far removed from British Rail's curly-edged sandwiches as could be imagined.

For the following week we were all scattered about the properties of the Reims members of the Academy and I, along with two others, was put in a palatial apartment above the offices of Lanson Père et Fils. This was delightful, with three bedrooms, one gold, one green, one blue, each with everything, including bathrobes, in the appropriate colour. There was a communal sitting room large enough for a worthwhile party and a servery which contained a fridge absolutely full of Lanson Black Label in bottles, magnums and even some of those Imperial Pints so beloved of Henry Longhurst.

These were for our use and we tucked into them whenever we came back to the flat, even though we were merely topping-up the system from all that we had at the other Houses during the week. The wonderful thing about the fridge was that it was always full and we never at any time saw anyone fill it – obviously magic.

Although I thoroughly enjoyed being my own man for the fortnight, rather than Gus's son, the Champagne region was not totally Gus-free during that time. I should have known that he would have to appear, and so it was in the second week, when he was driving with some friends to Strasbourg, unbelievably to visit the Appolinaris mineral water company. They stopped overnight in Reims and Gus decided to look me up that evening. He discovered that I was staying at the Lanson flat and duly arrived there at, quite a late hour. After knocking loudly on the front door in Boulevard Lundy, he enquired of the face that appeared at the second-floor window, "Is Alan Payne stopping here?" and was totally bemused by the response: "Yes, just leave him on the doorstep." I only heard about this on my return to London, as I never did see him in Reims.

At the end of the two weeks I arrived back in London with a very severe thirst. It seemed that on the plane there were a dozen young men with but one thought – a pint of English bitter. Champagne is wonderful, but two weeks of nothing but Champagne is seriously thirst inducing.

Having been on the trip, we were all invited to join the Old Boys of the Academy in the UK whose purpose was to enjoy drinking the wonderful product and to act as ambassadors wherever we could. A noble aim and, after what we had been treated to, how could we refuse, or indeed why?

So I attended the first annual dinner at the Dorchester where all members invite the principals from France and pay back some of the entertainment we enjoyed – although I hasten to add that they provided the Champagne. I duly joined the committee of the Academy in London and spent the next ten years as treasurer and secretary, which gave me my first taste of being involved in any organisation, however modest, outside of Paynes Caterers and away from Gus. That I thoroughly enjoyed it, and the opportunity to meet a range of people outside of the normal run of customers and colleagues who were my daily bread and butter, goes without saying. Involvement with the Academy was a very welcome change and it proved particularly pleasant when, a dozen years later as chairman, I was able to invite Gus as my guest to the annual dinner of an association in which he had no part.

During my year as chairman I had the arduous duty of presiding over a number of dinners around the country, including a four-day sojourn taking in banquets in Manchester, Edinburgh and Dublin. I have to say it was a very jolly task. Each dinner was organised by the local committee of the Academy so all I had to do was arrive, join in the fun and try to make a sensible speech at the end of the dinner.

Since this was twelve years after I had been on the Academy course, the full circle had gone round and André Rouzaud was once again the President of the Academy and he and I thoroughly enjoyed each other's company attending the three dinners. André, a very elegant understated man, was primarily a financier who also happened to run one of the world's finest brands - Roederer Cristal is seen by many as the best of the best. He stood out from the other proprietors who were Champagne producers through and through, most of them the latest of many generations of their families. It was always hard, given the glamour of their product, to remind oneself that by and large these were farmers. And, typical of farmers, they always spent a great deal of time complaining about the vagaries of the weather, which had meant the production of either too much or too little wine. Vineyard woes were never a subject I heard André talk about. On our tour round the regional dinners his main topic of conversation was his stables at Rouge Terre in Normandy, where he bred champion ponies for the very elegant trotting races that are so popular in France and America, but have never succeeded in England.

Since we were allowed to invite guests to the dinners, I decided to have as my guest at the Irish dinner a roguish, silver-tongued Irish friend, Pat Heneghan, who at that time ran the Irish Open Golf Tournament for his employers, WD Carroll, the Republic's largest tobacco company. Charm personified, I had known Pat for several years as he visited many tournaments in the UK developing contacts and enlisting players for his own event.

A phone call a couple of weeks before. "Pat, I have to chair a dinner for the Champagne Academy at the Shelburne Hotel on Saturday fortnight would you like to come as my guest?" "Delighted, dear boy, now when are you arriving?"

"On the Friday morning, straight from the Edinburgh dinner."

"That's good, because you'll need at least twenty-four hours' practice."

So I knew I was in for trouble. Pat was a world class act when it came to drinking. I had heard rumours that he had vowed never to touch a drop while he was in Ireland but I had only seen him in the UK and he certainly kept up a merry trot whenever he was there. What I didn't know was that Aer Lingus had instructions to give it to him intravenously as soon as the plane took off from Dublin.

André and I arrived at Dublin airport on the Friday morning and, while I was taken away to my fate by Pat, André was met by his agent in Ireland who was going to take him on a serious selling mission around the wine merchants of the City.

What followed for me was a day straight out of James Joyce, which started off very badly, when in the grand restaurant where we had lunch he asked me, "Now, as Chairman of the Champagne Academy, what's your favourite?"

I replied, "Probably Roederer Cristal."

"A magnum of Roederer Cristal, please" said Pat to the waiter.

When it arrived with one Champagne flute and a bottle of Perrier water for Pat, I knew that the rumours I had heard about Pat being teetotal in Ireland were true; I was in serious trouble as this magnum was just for me. Fortunately, a prodigious number of Pat's friends came up to talk to us, each of whom was offered a glass, and the magnum on its own was never going to be man enough for the task - luckily the restaurant appeared to have an inexhaustible supply. There followed an afternoon visiting what seemed like an endless Leopold Bloom parade of pubs full of Pat's friends enjoying a jolly preparation for the weekend and nobly helping me with my 'twenty-four hours practice'.

I was due to be joining André and his agent for dinner at the St Stephens Green Club, a rather stuffy, formal dining club of the sort I have steadfastly avoided in London. Pat decanted me at 7.30 at the door of this club with the words: "I'll put your luggage in the Shelburne and collect you at eight in the morning."

I staggered into the club and met up with André, his host and his host's wife, all of whom appeared not to have had a drink all day. I recall very little of the evening but according to André subsequently, I was on good form.

The following day is an autumnal blur, a non-stop tour of the countryside outside Dublin, visiting some very welcoming pubs and relishing seafood, brought ashore that morning and squeaky fresh. The day ended with Pat returning me to the hotel in time for a quick shower, into my dinner jacket and downstairs to meet the Irish branch of the Academy and their guests. The twenty-four hours practice had worked its charm and I had a cracking evening, which was summed up by the start of my speech, which set the tone for the long evening that followed, "May I introduce myself? I am Alan Payne and I am an English eedjit. No one but an English eedjit would come over here to lead the Irish Wine Trade in talking and drinking."

7

Further Steps

The process of getting away from Gus's all-pervading influence was helped enormously when I married Judy except, of course, for the wedding reception itself, which took place in my parents' garden in Acton. We had chosen, somewhat recklessly, to get married on 1st April.

I met Judy just before Christmas 1965 through my old school friend Peter Challands. I was immediately smitten by Judy, not knowing that she had observed me at Peter's twenty-first birthday party three weeks prior and thought me an absolute wally. She now says that nothing I have done over the intervening forty years has changed that impression. Heather, her flat-mate, persuaded her to accept my invitation to dinner by advising, "He's in catering, so at least you'll get a good meal." There are worse ways to start a relationship. When, on that first date, I discovered that she drank halves of bitter, I knew that Judy was the girl for me. Sadly I have weaned her on to more expensive beverages since.

Four months later, just before I was about to start my

seasonal peregrinations around the golf tournaments and agricultural shows for the whole of the summer, I proposed to her. I wanted her to experience the reality of my life as a peripatetic caterer for a full season before we got married. Not really having any idea what I was talking about, but bowled over by my charm and suavity, she accepted. So we became engaged and we planned to get married the following year, after which time she would only have herself to blame.

Judy was a lovely, petite, laughing lass from Lincolnshire, with, at that time, a fashionable 'beehive' of blond hair and, what delighted me as a 'bottom' man, was the very flattering ski pants she wore, made by the fashion company she worked for in the West End. It was only much later that I was to discover the true benefits I was going to get in Judy as a wife and partner in developing the business. Over the years she became more and more involved in helping me entertain our clients and many other guests, as our business was all about making people feel welcome. She was, quite brilliant at that. She could talk to anyone about anything and make that person feel very special. I'm sure that I got invited to many an event by a host not for my own merits but just so that he could sit Judy next to a difficult or shy guest, knowing that the guest would be looked after and happy – quite a gift.

Judy had grown up in a Methodist community in a market town called Alford, but she always knew the constraints of life in Alford were not for her and left home as soon as possible, first to study and then to work in Nottingham prior to moving to London.

It was normal at that time for the bride's family to stage the wedding but, by 1966, she had lived away from home for some seven years. Since all our friends were in London, and my family had the wherewithal for a jolly good shindig, we decided that London it was to be. Besides anything else, what on earth would the Methodists of Alford make of Gus? So then it was a question of convincing Judy's father, which we did by my obtaining tickets for the League Cup Final at Wembley and entertaining him to a great day out – shades of my future life selling corporate hospitality to companies wanting to sway opinions.

Naturally, Gus took charge of preparing the wedding reception. He also said that he would arrange for the coach and four grey horses, in which Rothmans, one of our most important customers, delivered their cigarettes around the West End, to take us from the church to the house. The week before the wedding he even went to the Easter Parade in Battersea Park where it was appearing to check that all was in order, or so he said. I was so used to Gus organising things that it never occurred to me to question him.

The day before the wedding when everything was set up, Gus arrived home from a very long lunch, very much the worse for wear. He lurched up the path accompanied by Andy Cotterill, he of the rotten scotch eggs, whom Gus had met in a club after lunch and who was very much responsible for Gus's current condition, which Andy shared. They staggered around the marquee in the garden, much to my mother's annoyance, and, before he finally left, a very drunk Andy persuaded Gus to lend him fifty pounds so he could give it to me as a wedding present. I was not unaccustomed to Gus and his cronies behaving badly and, in any case, I was more than a little preoccupied with my own nervousness the day before my wedding, but Lila was seething. She didn't speak to my father on the day itself or for weeks afterwards. Needless to say Gus never saw the fifty pounds again.

So we came out of the church on All-Fool's Day and, after the interminable delay over photographs, we made our way to the gate of the churchyard expecting to see the elegant Regency-style Coach & Four. More fools us! What greeted us was not a Coach & Four but Steptoe's Horse & Cart, with a couple of gilt chairs on the top. It was pulled by Hercules, the original horse from the television series, which, since it was one of the most popular programmes of the period, meant that our shock and mortification was witnessed by a gaggle of press cameramen, organised, naturally, by Gus.

After the reception and the guests had waved us away in my confetti-filled, tin-can-towing Triumph convertible, they all returned to the marquee fully intending to carry on where they had left off. However Gus had other ideas and he went on to the stage and through the mike said, "Right you've had all you're getting, time to bugger off." And as a friend reported, "So off we all buggered."

The evening was not as abruptly finished as that, as a party of twenty friends and family were treated to a very special dinner at one of Gus's West End watering holes, the Wellington Club, doubtless paid for with cases of Gordon's Gin. Judy and I meanwhile, having tidied up the car, arrived at the Branksome Court Hotel in Bournemouth (I was stylish even then) for four days of honeymoon, which was all I could get before starting my summer events season. When we arrived and requested dinner we were told that the dining room had been taken over for a function. However they could provide us with a 'cold collation' and set up a table behind the stage. So we spent our wedding evening, not being dined in splendour like our friends in London, but having a curly ham salad watching from the wings as the members of the Bournemouth Magic Circle performed for their colleagues.

At least after our marriage I had in Judy a very welcome and steadying counterbalance to Gus in my life. I was also able for the first time to take a more objective view of my relationship with Gus and started to get somewhat frustrated with the fact that I was still very much 'little Alan'. Gus's view inevitably prevailed and if I had the temerity to argue, a cigarette was put in my mouth and the argument was at an end.

During the sixties my responsibilities had expanded and I had created a wholesale wine business, primarily to purchase all our catering wine and spirit supplies at the correct price. It had however developed into a reasonable business itself, which involved me more and more in the wine trade that I had grown to enjoy following the Champagne Academy.

I had also helped to run the accounts department under the rather pompous John Marotto, the company's finance director and company secretary. Bald-headed and with round owl-like glasses, he was a timid version of Captain Mainwaring in Dad's Army. He would phone his bookie everyday and whisper his bets for the afternoon, all the time looking around to see if his wife was watching. Since she never came into the office it was obviously just paranoia or guilt. When he died very suddenly of a heart attack in 1966, I had to take over the entire accounts at the age of 22.

In spite of all this, by the end of the decade, Gus still treated me as 'Little Alan'; for example, I was not, after eight years in the business, a director, in spite of the fact that I was doing the finance director's job and running the wine business. He still expected me to collect him from his evenings out in the West End, sometimes waiting in the car until three or four o'clock in the morning when I had to decant him into the car and out again in Acton. Having done this ever since learning to drive, it was only when I was married to Judy that I realised that it was a strange thing for someone in his mid-twenties, about to become a father himself, to be doing.

Primarily I was very frustrated that I was not able to affect the direction the main catering business was taking, which I felt needed changing. It was, of course, I realised much later, a typical generation thing, one starting, one finishing, but at that time Judy and I seriously discussed whether we should emigrate to Canada. What stopped me was the fact that the company became involved in a couple of major events which Gus would have to let me run without interfering.

Gus had catered at the Farnborough Air Show since 1948. This was a showcase for the British Aviation industry and had become a caterers' benefit as several hundred companies showed their wares and entertained customers from around the world. A trade association called the Society of British Aerospace Companies, or SBAC, ran it, very successfully. The aviation industry was still, in the late nineteen-sixties, managed by men with handlebar moustaches who had won the war in their Lancasters and Spitfires. They had been celebrating in various bars ever since. Not surprisingly, Gus found this a very amenable environment and one in which he was very popular.

Most of our competitors were also involved and although we fought like fury to capture customers, once on site a mutual assistance operation existed. It was a marvellous opportunity to meet and in the evenings eat and drink with one's competitors. Initially it was an annual event but by the late sixties it had started to alternate with the Paris Air Show at Le Bourget.

At the end of the sixties, the SBAC created a British Pavilion at Le Bourget as a joint venture for all the British aviation companies who wished to exhibit their wares. This had a substantial catering facility and initially Ring & Brymer, our large rivals, did the catering for it. Their parent company, Forte, had just purchased three grand hotels in Paris, which R&B used as a base.

John Coomb, then a director of Rings, whom I had worked for at Ascot, decided that he would take a few friends over to the show in Paris and see how his French colleagues were doing. Naturally one of the rogues he took was Gus. According to John, they had an extraordinary couple of days on the loose in Paris. John was horrified by the amount of money the French hotels of Forte were charging to act as a remote base and decided that, since there was little he could do about it, he and his friends would sort out the Paris nightlife instead. This they did to the extent that Gus missed the plane back with all the others and so stopped another night.

The day he finally arrived back was the opening day of the Daks Golf Tournament, at that time the major summer event at Wentworth Golf Club near Virginia Water. I had organised the set-up of the tournament on my own and on the final preparation day I had expected Gus to arrive late afternoon and, as was his wont, change everything. I then got a call from John Coomb telling me that Gus had missed the plane and would no doubt be coming back next day.

The Thursday arrived, the event was set up and lunch was about to be served, when I wandered into Gus's tent and there he was, having crept in from the airport with no one seeing him, sitting quite still, with the only hangover I ever saw him suffer. It was half past eleven and he said, very slowly, “Open a bottle of Champagne please, quietly”. He then had one glass and went home.

Out of that trip, besides allowing me to set up and run a tournament without Gus altering everything, came our involvement for the next ten years with the Air Shows in Hanover and Paris, and my chance to prove myself, as Gus didn’t speak French or German. Mind you at that stage, neither could I with any fluency – while sober that is. The difference was that firstly, I had learnt both languages at school, and I had ‘done a vintage’, picking grapes, in both countries, in Bordeaux and on the Rhine, during which time I had spoken no English at all. I also was prepared to learn whereas Gus, like most Englishmen of his generation, merely increased the volume in English.

Finding suppliers and staff in Paris was extremely hazardous, fraught and testing, but vastly entertaining. One of Gus’s friends, Max Joseph, had, like Charles Forte, also bought three trophy hotels in the centre of Paris and Gus arranged with him that the chef in one of them, the Hotel Scribe, would cook meat for us. Given what Forte’s hotels had done to Rings, I was nervous about this and was right to be. In spite of the vast back-hander we had to give to the chef before he would talk to us, he refused point-blank to cook the beef any other way than ‘bleu’, in other words virtually raw, which on a buffet for mainly English customers was totally unacceptable. So we cancelled him after one collection.

Luckily I had established several alternative sources, although only after some near disasters. On my first preparation trip to Paris I had been given the name of a general supplier in the new market at Rungis, south of the city. This market had recently replaced the old city market of Les Halles, which I had visited as a thirteen-year-old on the school exchange trip in 1958.

I went by taxi to the massive new market south of Orly Airport and after a highly confusing tour of several of the vast warehouses I eventually met this man, a Monsieur Porta. He was a large jovial chap with a very red face, who spoke no English, but who, when he heard what I was there for, sat me down at a table, took a chair opposite me, pushed several piles of paper to the side and reached for a bottle of Pastis (thankfully not full) and two glasses to aid the discussion. I was enormously relieved that he appeared able to source most of the things I needed in the way of food and dry goods, leaving aside the beef that I had believed I was going to get from the Hotel Scribe.

I left after the bottle was finished, secure in the fact that M. Porta would write to me with the definitive prices for everything. I waited and waited but no letter was forthcoming. I tried telephoning, three times with no success; I imagined it ringing in the cluttered little office in the corner of the gigantic depot. On the fourth attempt the phone was answered by a female voice and, being somewhat less adept at French on the phone, I asked,

"Puis-je parler avec Monsieur Porta?" ('May I speak to Monsieur Porta?')

"Non."	('No')
"Pourquoi pas?"	('Why Not ?')
"Parce qu'il est mort."	('Because he is dead.')

This was not helpful. Firstly, I was obviously talking to his wife and I was not very good at condolence phrases in French. Secondly, I now had to find a new supplier if she couldn't help, which in the circumstances, I thought unlikely.

In the end it meant another trip to Paris and I found a couple of suppliers, one within a couple of miles of Le Bourget, and the other, a truly wonderful traiteur (the French word for caterer, which also means delicatessen), in the Rue St Honoré in the centre of town, who appeared to be able to magic supplies out of thin air, including perfectly cooked beef when the Scribe let us down.

However, the one thing we couldn't get in France was Stilton cheese, which was, of course, absolutely vital for the continued wellbeing of our English customers. There was no way that we could buy it in France or indeed legally import it. We therefore decided to smuggle it in. Altogether we had fourteen whole Stiltons scattered between the four cars that we used to transport some of the staff. The only car stopped at Calais was Gus in his Rover. The customs officer asked Gus to get out of the car and looked at him in absolute astonishment. This was not really surprising, as he was dressed in a green tweed Norfolk jacket with plus-fours, brilliantly polished brown brogues, a gold chain with an Albert watch spread across his ample belly, a red carnation, mutton-chop whiskers and a cigar; he looked every inch an Englishman from central-casting.

The customs officer instructed Gus to open the car boot. Inside were two suitcases and eight cylindrical containers. Four contained Stiltons and four held Gus's hats from Locks of St James's.

"Qu'est-ce que c'est?" he asked, pointing to the containers.

With enormous aplomb, Gus opened one of the hat boxes took out a brown bowler, put it on his head, tapped the crown and said very loudly and with an atrocious accent, while pointing at the bowler, "Chapeau!"

The douanier, and everyone else in the customs shed,

collapsed with laughter and Gus, together with the Stiltons, was waved through.

Having realised that he had to let me get on with it, Gus's role in Paris became very much that of the grand seigneur. He would arrive at Le Bourget around midday, having driven his very large Rover from the centre of Paris with magnificent disregard for all road signs and rights of way and an admirably total distain for the lethal traffic on the dreaded Boulevard Périphérique around Paris.

He would then, after a glass of Champagne, set himself up on his shooting stick, cigar in hand, at the start of our very elaborate buffet, collecting the lunch tickets, which were our currency. This would, of course, give him the opportunity to chat with all his pals as they queued up for their lunch.

We always had a considerable choice on offer, which was very necessary since most of our customers were there for the entire ten days. In spite of this, some of the clients were never satisfied; however some regretted raising the subject with a harmless question such as, "What have you got new today, Gus?" which inevitably produced the waspish come-back, "Why? Did you have all sixteen choices yesterday?"

The air shows themselves were always hard work but hugely agreeable, enormous fun, and it was a great joy having Gus there for the first few years. He was in his element just taking pleasure in being with his old Farnborough cronies, many of whom, while very important business figures, were as delightfully anachronistic and eccentric as he was.

One of the most outstanding was in fact in charge of the whole British contingent in Paris, the Director of SBAC, Admiral Sir Richard Smeeton. Dick Smeeton was a massive man, six foot six and twenty stone, with a booming voice that must have carried across the decks of various fighting ships over the years. My particular memory of Sir Richard was in the Director's office. He had summoned in a pressman who was, to put it kindly, with wine. As a newt. He was put in a chair opposite Sir Richard, and given a dressing down that everyone in the British pavilion, if not the entire show, heard. As, trembling, the journalist got up, he knocked over the table and emptied a jug of hot coffee over Sir Richard who exploded with rage.

We just about got the man out alive, and I was helping the Director recover himself when he informed me, "Understand, Alan, that man is never to be served here again! Understand, never!" After I nodded he continued, "Can't have people drunk on duty." I nodded again. "I wouldn't have minded, but he burnt me balls – and I don't like having me balls burnt!" No argument from me.

As the show went on for ten days, we were away for over a fortnight, longer than our normal events. That first year was more than just a debut for me; I broke protocol and flew back for a day during the first weekend to see Judy, who was returning home from Queen Charlotte's, having given birth to our son Richard just before I went to Paris. (It was lucky he arrived when he did as I warned Judy that I had arranged with Hawker Siddeley, the aircraft manufacturer, that if he arrived during the show they would fly me back in a Harrier and land on the hospital roof – to the accompaniment, doubtless, of hundreds of early births.)

During the following show, Judy came out to Paris for a few days with her great friend Mo Dixon, from her Nottingham days. Mo was a very formidable lady, who spent her whole life overcoming the male chauvinism that was the Yorkshire of her youth, succeeding in becoming a very senior figure in health service administration. Her staunchly held, and robustly expressed, left wing views clashed frequently with my own, as my liberalism tended to be more that of the nineteenth rather than twentieth century variety.

Judy and Mo had a great time, being guided around Paris by Mo's old pen friend Colette, a university lecturer at the Sorbonne. Colette lived in the smallest flat I have ever seen, with the loo housed in a cupboard under the stairs and the shower in a half-metre square alcove off the kitchen. The three of us went there for dinner with Colette and a female friend of hers – myself and four ladies, two French, two English. Judy spoke very little French, Colette's friend no English at all, so I spent the evening hilariously trying to translate English jokes into French – my translations were infinitely funnier than the jokes.

We had a beef fondue and I remarked that the rule was that if I allowed my meat to fall in the fat, I had to kiss everyone. Trying to translate that, I used the French word 'baiser'. Innocent that I was, I had no idea that 'baiser', while it does mean 'to kiss', also colloquially and more robustly is used to mean making love. Mo, Colette and her friend, all French speakers, quite naturally fell about with shrieks of Gallic laughter; saying in equally vigorous French that they hoped the Englishman was up to the task. Needless to say I didn't allow my meat to fall.

In between each show I made a habit of visiting Paris for one or two preparation trips, meeting up with my suppliers, sometimes finding new ones and thoroughly enjoying myself. I could, of course, have done it all by letter but it was the only indulgence I allowed myself, and it was very agreeable and rather good for my self-esteem, bearing in mind that this was during the seventies, a very grey decade in London when elegance and luxury were political no-no's.

One of the most outstanding of my contacts during that period, and one who seemed to enjoy my company as much as I did his, was a very aristocratic Frenchman called Count Philippe de Lafayette. Philippe was the descendant of the great General and, since he had no offspring, was probably the last of the line. He worked for Mumm Champagne as a sort of super-rep with a particular interest in making certain that Mumm was the Champagne that the aviation industry considered its own, hence his initial interest in my operation at the air show at Le Bourget.

Mumm Champagne was, at that time, owned by the vast American drinks company, Seagram and, as the last of the Lafayettes, Philippe was used by Seagram to entertain visiting bigwigs from the States with enormous effect. This meant that he had an absolutely limitless expense account and was treated as a god by every hotelier and restaurateur in Paris. Over the years he took me to some of the most outrageously swish, and doubtless hideously expensive, places. These were lovely occasions but mostly I will be forever grateful to Philippe because he gave me the opportunity for the only time in my life to upstage a seriously snotty French wine waiter.

This was just before our fifth Le Bourget show and the company had grown to a size where it could not afford the luxury of me being away for a fortnight and for all the preparatory visits. To my chagrin, I decided I had better take one of our catering managers with me in order that he could take over the following year. Thus it was that Jon Davies and I arrived at my usual hotel in the Avenue Villiers to be greeted warmly by the lady owner with, as was normal in those pre-mobile phone days, a host of messages awaiting my arrival. She was very excited because Monsieur le Comte Philippe de

Lafayette had phoned three times. The French bourgeoisie, while it might be quite keen on Liberté and Fraternité is not very hot on Egalité. I fondly remember over breakfast hearing Madame, on the morning after one of Philippe's visits to collect me, saying very grandly to her cleaner that Monsieur le Comte Philippe de Lafayette had been in her hotel the previous day and that he had kissed her hand.

On this occasion, I phoned his secretary who said that unfortunately Philippe had to go to a funeral in the south of France the following day and would not be able to dine with me as we had arranged. The only possibility was that very evening before he caught the night train. I said that I had already arranged for Jon and me to have dinner with a friend, Colette, and perhaps Philippe could join the three of us. The secretary said that Philippe would insist on hosting us and she was sure that Ledoyen would be able to accomodate the four of us. So it was that Philippe, to the absolute delight of the hotel patronne, collected us from my hotel at seven thirty.

Since Colette was very much a latter-day Simone de Beauvoir, with long dark straight hair and a penchant for the music of Leonard Cohen, whose natural habitat was the bistros of the left bank, I was not sure how she would take to the elitism of Ledoyen, a remarkable restaurant housed in a very grand nineteenth century detached villa. Set back in the woods off the Champs-Elysées at the lower end, towards the Place de la Concorde, it is swank in the true pink-tablecloth tradition of Michelin-starred restaurants. I needn't have worried, as the French would never allow political prejudice to interfere with a fine meal. (Having said that, in thirty-plus years I never came across a British socialist who objected to enjoying the privilege inherent in any of our events.)

Ledoyen was truly Philippe's home from home. Philippe didn't park his car, merely shrugged it off like a slipper. He stepped out leaving the engine running secure in the knowledge that it would be dealt with. With the ceremony due to the 'Roi Soleil', we were ushered in and were taken serenely through to his table, already graced with a bottle of the current vintage of Mumm Champagne in a vast silver bucket. As we enjoyed the Champagne, Philippe started the process of discussing with the proprietor the dishes and the wines that would be suitable for his guests, without worrying us with the troublesome details.

I had by this time known Philippe for some ten years and he had developed the engaging habit of pushing the first sample of wine, given to him by the sommelier, over to me to taste rather than do it himself. It was a rather flattering gesture which I found most beguiling. However on this particular evening, the sommelier was obviously not as anglophile as Philippe, and took considerable offence that the 'jeune Anglais' and not 'Monsieur le Comte' was passing judgement on his wine.

After the first two wines had been passed over to me, with a brief "Goûtez, Alain" from Philippe, and each time I had nodded approval, the wine waiter was seriously agitated and not a happy French bunny at all. The third wine was a claret served from a basket and the sommelier, with his hand over the label of the bottle covering the wine and the basket, duly served a drop for Philippe to taste and Philippe slid the glass across the table to me.

By this time the waiter was behind Colette, opposite me, waiting to serve her, when I said, "Qu'est le Bordeaux?" (What is the Bordeaux ?) The waiter virtually spat out the words through curled Gallic lips, "Pichon-Longueville", and, as a very,very belated afterthought, "Monsieur." It was quite impressive the venom, obviously the culmination of forty years of dealing with perfidious Albion, that he managed to get into that "Monsieur".

Now it just so happened that at that time there were two Châteaux called Pichon-Longueville: Château Pichon-Longueville Baron and Château Pichon-Longueville Lalande. The names have since been changed to avoid the very confusion of which I was going to take full advantage, "Ah oui, Baron ou Lalande?" (Ah yes, Baron or Lalande?) And he didn't know.

A lifetime passed for him - and then he had to look down at the bottle - and, surreptitiously, move his hand, so he could see the label at which point he very shamefecedly said, "C'est le Baron, Monsieur." (It's the Baron, sir.) Graciously I allowed him to go ahead with the words, "Pour le Baron, c'est pas mal." (For the Baron, it's not bad.) After the sommellier had finished serving the claret, Colette, the class warrior, who had, of course, been watching the entire battle, leaned across the table, clapped her hands and said, "C'etait manifique ça!" (That was marvellous!)

It was an exquisite moment and though I saw the waiter on several subequent visits he was never rebellious again. It is nice to know that I did some good.

What a shame Gus wasn't there – he wouldn't have understood it all – but he would have loved it nonetheless.

8

Surely They Played Liszt

Gus was such a life force that it seemed as though he was always going to be around; however, given his lifestyle, this was never a likelihood. He ate and drank far too much for most of his life and, from the war onwards, he did remarkably little exercise except for his right arm, which was certainly well used and so kept the rest of him well oiled.

He also chose, in function catering, one of the most stressful of jobs, totally time-critical and, most frustratingly of all, one largely dependent on other people being able to fulfil their promise to you so that you can deliver yours to the customer. In addition, particularly in the early days of the business, he was always breaking new ground in what he offered to visitors to events. Most of these new ideas were tried without any dress rehearsals, instantly either a success or failure, in the full glare of the client's expectations.

The first sign of his mortality was in the early nineteen sixties. He had a pain in his chest and Wilf Winder, his doctor, whose daughter's wedding had started Gus's catering career (and cost him his car seat), sent him to see a consultant. I drove him from our office in Shepherds Bush to the consulting rooms in Montagu Square - not his usual trip to the West End. I waited outside and a very subdued Gus emerged an hour or so later with the instruction, "To Harrods"; when I asked why he said, "He told me I've had a heart attack and if I've got to be in bed for six weeks, I'm going to be bloody comfortable. The doctor had a wonderful cashmere blanket, which he got from Harrods so we're going there to get one."

Gus held court in his bedroom, with his cashmere blanket, for the next couple of months and, besides the numerous visitors that my mother had to cope with, there was always racing on television. On one day he bet on forty horses and ended up sixpence ahead of the bookies.

The only subsequent alteration to his lifestyle was that he changed his tipple from gin to scotch. This was on the specialist's advice as, apparently, gin thickens the blood and whisky thins it. The tale went round the trade that Booth's had shut one of their stills when they heard the news of Gus's quite literal change of heart, and cheers went up in the Highlands of Scotland.

The one friend who cheered the most was Robin Cater, at that time Managing Director of John Haig, makers of the famous Dimple Haig whisky and Gus's new tipple-of-choice. After Gus was better and back to his normal antics, he invited Robin to a day at the Daks Golf Tournament at Wentworth. Gus looked after him famously, told him where to go on the course to find the action, when he should be back at the tent for 'elevenses', and then for lunch and tea. He gave him an excellent time at the end of which Robin in his dry Scottish way said that he had so much enjoyed the day that he thought he might very well take up the game himself. Gus was so pleased with this comment that he subsequently sent Robin a dozen golf balls made by his friends at Dunlop, with a note saying that he hoped Robin enjoyed his new sport.

Robin was very touched by this unexpected and generous gesture. However, what Gus didn't know, and Robin had studiously avoided mentioning, was that, far from not knowing about golf, Robin had once been the captain of the British team for the Walker Cup. This is a biennial match between teams of amateur golfers representing Great Britain and America – the amateur Ryder Cup. It was some time before Gus found out about Robin's subterfuge, and when he did, it naturally caused a typical Gus guffaw at the fact that he had been taken in.

Gus's generosity was on this occasion handsomely rewarded some eight years later. By the early seventies, Robin had become the Chairman of the Championship Committee of the Royal & Ancient Golf Club of St Andrews and, since one of the events which the R&A ran was the Open Championship, Robin Cater and Keith McKenzie, the Secretary of the R&A, wanted Gus to take over the catering for it.

For as long as anyone could remember, the Open, the biggest golf event in the world, had always had one caterer undertaking the whole event. However, since at that time the championship took place each year at a different seaside 'links' course on a roster of primarily seven golf clubs, three in the north west of England and four in Scotland, it was the bizarre habit of the R&A to change caterers when the championship moved. This gave the Scottish championships to a caterer in Edinburgh and the English events to a firm from Manchester, an arrangement that inevitably caused some problems with continuity and ignored the strengths and weaknesses of each caterer.

Gus, as the caterer for somewhere between sixty and seventy percent of the professional tournaments of the PGA for the previous decade and a half, had frequently been asked to cater for the Open but had refused, as he thought it was too big for him to take over in its entirety. What Robin Cater and Keith McKenzie were prepared to accept for the first time, was that Paynes Caterers could develop and operate the hospitality pavilions, for the many companies who wanted to entertain their guests at the Open, and the existing contractors could alternate, as before, the running of the public and official catering.

In the summer of 1972 it was agreed in principle that this change would take place the following year at Royal Troon and consequently Gus, Alan Cox and I went to the championships at Muirfield in July 1972 to study the event in detail and put together a report for the Championship Committee on how we felt the event should be run from a catering and hospitality point of view.

Although he hadn't catered at the Open before, Gus was nonetheless a regular visitor to the Championships as it was, after all, the event to be at for everyone involved in golf. In 1969 there had been an addition to the facilities which made him feel very much more at home, when Anthony Leschalles, the shipper of Champagne Bollinger, decided that there was a market for a Champagne tent at the Open, (doubtless spurred by the fact that Gus's 'elevenses' had made Champagne the drink of choice for 'those in the know' in golf).

It was an inspired decision, helped enormously by the fact that a British golfer, Tony Jacklin, won that year's event, which caused a considerable increase in sales - Anthony was, however, somewhat miffed when he presented a magnum of Bollinger to Tony on his victory, only to be told by the already ubiquitous Mark McCormack that it would only be opened by Tony on payment of £500 - a sign of things to come.

Gus naturally used the Bollinger Tent as his Open Championship base from 1969 at Lytham onwards, (having, of course, done a deal with Anthony over sensible prices and methods of payment), and he held court there every day he was at the Open. It was a wonderful seeding of interest for Bollinger, particularly as it gave an opportunity for the more genuine of Gus's friends to buy the odd bottle or two and repay the many favours Gus had given them over the years. Subsequently the 'Bollinger' has become the definitive meeting place at the Open Championships and thirty-six years later is still the pre-broadcast haunt of Peter Alliss, his wife, Jackie, and a very large merry band of the usual suspects.

A few years ago the ever-amusing columnist from the Daily Mail, Ian Wooldridge, wrote a lovely piece about how, as a journalist arriving at the Open, it was necessary, in order to prove one's professionalism, to personally measure out the important yardages of the course each year. As a consequence, he was able to say, having measured the distance personally, that at that particular event, it was exactly two hundred and ten yards from the Press Tent to the Bollinger Tent.

So, when the three of us arrived at Muirfield on our reconnaissance mission, Gus went straight to the 'Bollinger' and Big Alan and Little Alan went off to observe and report back to him a couple of hours later, by which time he had probably learnt far more from the people who came in to have a drink with him than we had by looking round.

What we saw was that the hospitality pavilions at the

Open had no order to them whatsoever; there were twenty to thirty marquees of varying sizes, which looked as if they had been dropped willy-nilly from a plane and were therefore scattered in a random fashion without any reference to how they were to be serviced. I have this abiding memory of a hapless young waitress carrying a tray of prawn cocktails from the kitchen marquee and wandering about, totally lost, looking for the tent where she was working and where, doubtless, some of the guests were keen to start their lunch.

It didn't take much to put the whole thing into a sensible formation the next year with access to all the pavilions from a central compound, which radically transformed the service to be offered. Once this was done, the hospitality pavilions at the Open became incredibly popular and an important source of income for the R&A.

It became so important that after the 1975 Open at Carnoustie, which had not been a financial success in other areas, the outstanding popularity and income generated by the hospitality operation made the R&A want to increase the numbers the following year at Royal Birkdale in Lancashire. So the Secretary, Keith McKenzie and his finance secretary, George Wilson, came down to see Gus and me at Shepherds Bush.

What they said excited and frightened us. They wanted to more than double the number of entertainment packages on offer in 1976. Totally out of character, Gus was very quiet throughout the meeting and let me do the talking. I said that what they were now proposing was not a planned growth but an explosion, and you cannot control an explosion. I proposed that we allow our units to grow by a third and that we set up a complementary hospitality operation to be run by one of the other caterers, but the same one in Scotland and England, to maintain continuity. This was agreed and we designed and organised the second operation and helped our great rivals Ring & Brymer to get the thing started. This arrangement worked well and has continued for the subsequent thirty years.

It was as a result of this meeting that Gus announced that I would become managing director of the company because he felt I had proved that I was not prepared to compromise his high standards for an immediate quick profit. I have to admit that I heard this reason from someone else, because he would never have told me himself.

Gus didn't take life any easier after his early heart troubles, or let up on his lifestyle, certainly not until the early nineteen seventies when, in 1973 after he had been forced to allow me to run the air shows in Paris and Hanover, he made me a director of the company that year and then in 1976 I was elevated to managing director. After that, I used to say only half-jokingly, he came in to work only once a week, gave me a bollocking, and went home.

At about this time we were approached by another delightful character destined to play a very significant role in the development of Paynes Caterers, Richard Goddard, the newly installed Secretary of Henley Royal Regatta. Richard discovered, when he arrived at the Regatta in the autumn of 1975, that the coffers were almost bare, as the event had never been sponsored, which remained a forbidden route to financial salvation, and, by the nature of the beast, the races took place on a mile and a half of river where the public could freely watch without payment.

One of his suppliers, Ron Millward, who erected the Regatta grandstands, told Richard that Gus and Alan Payne had started a hospitality operation at the Open Championships, where Ron also built the stands. He knew that these had been extremely successful and made a lot of money for the R&A as a consequence.

Richard, who, with his moustache and similar build to Gus, looked like a young Oliver Hardy, lost no time in coming to see us at Shepherds Bush to ask if we would be interested in trying out a hospitality operation on some riverside land the Regatta owned on Fawley Meadows.

Alan Cox and I went down to Henley, which I hadn't been to since working there in the early sixties for John Gardner, the caterers, subsequently one half of Gardner-Merchant and fated to become yet another part of the Forte Empire. At that time, in the 'Holy of Holies' of the Regatta, the Stewards' Enclosure lunch tent, everyone had lobster for lunch, some twelve hundred a day, and I remember spending four days cracking lobster shells for eight hours a day, with red raw hands to show for it on the Sunday evening.

My recollections of Henley, therefore, being less than Arcadian, I was still totally captivated by the riverside site that Richard proposed that we use for a hospitality facility during the Regatta week. Besides being a beautiful stretch of the Thames valley, it had the benefit of being able to grow almost endlessly, should we be successful, since the Stewards of the Regatta owned at least half a mile of riverbank, which, as it turned out, was just as well.

We agreed that we would put up a trial pavilion on Fawley Meadows for the duration of the Regatta that summer and invite potential customers along to try it out. We cobbled together a coalition of suppliers who were prepared to put up tents, provide furniture, wine, champagne, Pimms and other produce at little or no cost on the promise of some years as sole providers should the operation be a success.

We had a very enjoyable few days entertaining dozens of existing and potential customers and I was delighted that all the suppliers benefited greatly, as did we, as Paynes Pavilions at Fawley Meadows became one of the great successes of the embryonic corporate hospitality industry.

Gus had been very sceptical about the whole idea, but was persuaded to dress up and come along and join in the fun. I even went to Locks, his hatters in St James and, without his knowledge, had straw boaters made for us both to wear. He thoroughly enjoyed his day and grudgingly agreed that it might just work.

When, within a couple of years, the operation started to make more money than any event he had ever undertaken, he freely admitted that he had been wrong. Sadly the real growth was in the decade after he was no longer with us.

What Gus enjoyed as much as I did, besides the profit, was getting to know Richard Goddard and the new Chairman of the Regatta, Peter Coni, who was a QC and who delighted us with many wonderfully indiscreet stories from the Old Bailey. The two of them were engaging companions and came to several of our events and also participated in Gus's annual winter lunch held on the day after the PGA Dinner when most of the rogues of the golf world were in London. Peter and Richard were very happy to become honorary golf rogues once a year.

This was a very merry affair, held for many years in our cellars at Pennard Road, and subsequently at a wine merchant's cellars in the West End, and then at the magnificent Martini Terrace at the top of New Zealand House on the corner of Pall Mall and Haymarket with breathtaking views over central London.

The other annual lunch Gus arranged was for the secretaries of all the main people he did business with. Betty, Gus's secretary, would host this with Connie and the other Betty, our head waitress, and Gus and I would wait on. What this actually meant was that I waited on and Gus sat on a shooting stick by the side of the table smoking his cigar. It was invariably a great success and meant that Gus always enjoyed immediate access to their bosses and their bosses' diaries.

About the time he handed over day-to-day control of the company to me, Gus, probably feeling the effects of a lifetime's dissolution, came under the influence of a strange and very beautiful Indian lady therapist who put him on a very strict and bizarre diet with no alcohol and some very pungent oils for him to bathe in. He lost a dramatic amount of weight and looked pretty ghastly and I told him that he might be much healthier, although he certainly didn't look it, but he was a bloody awful advert for the company and its product. Worst of all, the wonderful and expensive suits made by Mr Pedersen, using the benefit of generations of tailoring skills specifically to disguise Gus's once portly shape, all now hung on him dreadfully.

The oils he used in his bath were unbelievably powerful and on one occasion when we were catering for a tournament at Gleneagles, he was staying at the hotel and I went over to see him in his room the first morning. The moment I walked into the very grand entrance to the massive hotel, I could smell the essences of his bath oils permeating the entire building, with all the tournament organisers, players and guests wondering what the hell it was. Needless to say I didn't enlighten them.

At the beginning of 1978 he had another heart attack, which could have been triggered by a rather difficult interview we had had with the local director of our bank. We went to see them to talk about purchasing new premises we had found in Chiswick in the charmingly named British Grove. It was an old printing works with a separate house and a yard that would easily accommodate all our vans and a few cars. While it needed some work done on it, it was perfect for our growing business, because our existing two properties in Shepherds Bush had become very difficult to operate as we were spending most of our time driving vans and cars around the block ahead of the traffic wardens.

Since we had never sought finance from anyone other than our local Barclays Bank, it seemed natural to seek their advice and assistance, especially since the local director had been Gus's guest on many occasions at events, and as such was considered a useful friend.

What is it about a bank that makes them take the umbrella back when it starts raining? We told them of our plans and showed them the growing order book and explained our current operational problems and hoped that, if they did not feel they wanted to lend us the mortgage on the property, they could give us some recommendation on who could. Not a chance.

A categorical, "You shouldn't do this," was all we got - no advice, no help. Gus's final comment as we left the office in Chelsea was: "We're going to buy it." And plaintively he said to me as we got out of the lift, "We are, aren't we?" We did, and although the purchase and the considerable growth we were about to experience, made possible by the move, was going to give us problems a couple of years down the line, these sadly were not going to be Gus's problems.

He once again suffered the chest pains and knew what it was this time. A visit to his consultant in Montagu Square confirmed it and on this occasion it was hospital not home rest. He seemed to be going downhill for the next couple of months and then, when the summer events season started, he rallied, to everyone's total astonishment and delight. He came out of hospital and went to the Martini Golf

Tournament at the RAC Club at Epsom to the extreme pleasure of his many great friends at Martini & Rossi who were one of his oldest customers.

It was while driving him home from this event that I had one of the strangest conversations with him. The following week we were due to be going to Royal Birkdale Golf Club on the Lancashire coast for the Penfold PGA Championship and Gus suddenly said to me, "What accommodation have you got for me at Birkdale?"

Considering that the week before he had still been in hospital, this was quite a question. My response was up to the task, "When I was doing the arrangements for Birkdale, you were in the process of dying, so I thought it not worth booking anything for you. Now that you've decided to live I suppose I'll have to organise something." He roared with laughter and said, "Bloody right."

So he came to Birkdale and then to St Andrews for his last Open Championship and the great thing was that he was his old self, none of the nonsensical diet and drinking water, he was enjoying himself thoroughly, eating and drinking as of old. He even hosted his traditional grouse dinner in August at the Mount Royal Hotel in York during the Benson & Hedges Tournament as he had the previous dozen or so years. No one could believe that he had been on death's door a few months before, especially when he and Lila, my mother, flew over to Dublin for the Irish Open as guests of the organiser, the irrepressible Pat Heneghan.

In short he visited all the events during the summer. However, when the season was over he crumpled just as suddenly as he had burst back to life earlier, having recently attended the funerals of his two great friends Henry Longhurst and Len Jolly.

A week after Tommy had taken him back to London

from the last tournament of the year, he was taken into Kingston Hospital. We managed to get him a private room and we set him up so that all the very many friends who wanted to pay their respects could enjoy a last glass of bubbly with Gus. He died the following week and it was only on reading the death certificate that we discovered that it was liver cancer, not his heart, that was the final killer. No one had ever mentioned cancer to any of the family.

It was good that he went quickly, as he would certainly not have aged gracefully, or been a stoic invalid, although as my mother said, "The silence after Gus is very, very loud."

A suitable farewell to Gus took a little organising. Firstly there was no way of knowing how many would come. Gus had touched so many and such varied lives in his time that it could be hundreds or even thousands. Besides all the friends and business contacts we knew, which ran into many hundreds, we started to hear, after the announcement, from army colleagues and people who worked or played tennis with him at Lyons and from his Shepherds Bush swimming days, as well as casual drinking pals from rugby and West End drinking clubs. There were also, of course, the many members of staff at the numerous haunts of his and to whom he was a generous and charming customer.

We did the best we could and went for the largest crematorium in West London. We insisted on a time for the service that would allow everyone who wanted to, and certainly all those 'in the know', to get back to my house in East Sheen by the right time for a final 'elevenses'.

I then had a bizarre discussion with the undertaker who couldn't quite grasp what I was saying when I demanded that we take the previous time slot at the crematorium as well as the one when the service would take place. I tried to get across that a great many people would be coming in from all over the UK as well as from abroad to say farewell to Gus. Besides making sure that they had somewhere to go, I did not think it was fair to have several hundred of them waiting outside, blocking the roads, while the poor mourners from the previous funeral were trying to pay a respectful farewell to their departed relative.

The undertaker said it could be done, although he had never known it before, but it would cost an extra seven pounds fifty. I agreed that for Gus we would lash out and to hell with the expense. The crematorium had a minstrels' gallery and I arranged for a quintet to play before, during and after the ceremony with the highlight being the wonderfully melancholy Brahms' Clarinet Quintet as we departed.

One of Gus's best friends, Bill Graham, who created the wonderful golf and country club, St Pierre, at Chepstow, the home of the Dunlop Masters for several years, was living at that time in the Isle of Man and he and his wife, Betty, stayed with us the night before the funeral. He told me how much Gus thought of me, how proud he was of me and of what I was doing to the company. I was completely unmanned by this, as I had no idea that Gus thought so much of me since he never expressed it. We Englishmen with our repressed emotions!

It was then my turn to have Bill in tears when I asked him to become chairman of Paynes Caterers in Gus's place. He agreed and was a wonderful figurehead and a wise advisor during a roller coaster few years. It was an emotional evening with many laughs recalling Gus's many antics as well as the inevitable tears.

When we arrived at the crematorium, the funeral procession had to thread its way through a great many people who couldn't get into the building and although the whole day is somewhat vague, my two outstanding memories are, firstly, as I walked into the porch I saw, standing together outside, the incongruous pairing of the vast six-foot-six form of Sir Richard Smeeton next to the five-foot-four Cliff Morgan. My other memory is after the service, of which I remember absolutely nothing, hearing the plaintive strains of the clarinet arpeggios from Brahms as we were leaving the building. Quite haunting.

Gus would have been very happy with the fact that we all arrived, (well, luckily not all), in perfect time for 'elevenses'. It was a truly joyful occasion, as befitted a man who never ever felt sorry for himself, always worried about others and made sure they were having a good time.

The frenetic activity of sorting everything out left me no time to consider how Gus's death had affected me personally but once the wistful jubilation of the funeral was over, I started to realise just what my mother meant about the loudness of the silence. Gus had been such an immense presence, and such a lovely one, that the gap was enormous and unlikely to be filled. Thirty years later it still hasn't.

There was a fitting corollary a few weeks after the funeral, while I was picking up the threads of a very changed business life. I was taken to lunch by an old pal of Gus's called Ray Watkins, a director of the company that made Gus's early-life tipple, Gordon's Gin.

Ray said to me, "Alan, I hope you don't mind me telling you this as it is a bit disrespectful, but I think you'll enjoy it. After the funeral the three of us from Gordon's went back to the distillery for a board meeting. Naturally we were asked how the funeral went and I said, 'It was a fitting goodbye to Gus, there were the best part of a thousand there and there must have been at least four hundred from the trade, and, a nice touch, they had a clarinet quintet playing Brahms.' Well, you know what a waspish fellow Colin

Ashmore can be, he suddenly said, 'For Gus? Surely they played Liszt as well.'"

A well-nigh perfect epitaph.

9

--- & GUNTER

Another of the mourners at Gus's funeral, if such a term can be used on such a jolly occasion, was John Coomb. As Gus's old sparring partner at Farnborough Air Show and numerous other watering places as well as my mentor at Ascot, Goodwood and the City, John was naturally keen to pay his respects and to have a few glasses with the old man for the last time. As he left, he said that I must come and have lunch with him at the Parlour at Ring & Brymer's headquarters in Sun Street in the City.

Good as his word an invitation arrived within a few days and I found myself in the dining room where numerous Lord Mayors and other City dignitaries had been entertained and beguiled into spending considerable sums of money on the very elaborate banquets that were Rings' captive markets.

As Sir Lindsay Ring had been Lord Mayor himself the previous year, this had reinforced their already dominant position in the City and by the time I lunched with John they were undertaking the vast majority of City functions.

After lunch when our dining companions had left, over the port, a fateful conversation started, "You realise, John, that since I've now purchased the new warehouse in British Grove, which I'm going to have to pay for, I'm going to have to attack new markets, most especially your bailiwick of the City. Thanks for the lunch, but just to let you know, I'm coming after you."

John chuckled, " Who are you going to have running the City for you? You won't have time yourself, and Alan Cox is not cut out for it." Spoofing like mad, I responded, "I have a few ideas but nothing concrete yet, but anyway, we're not moving into British Grove for a few months."

He then astonished me with the question, "Well what are you going to offer me?" "You're joking!" "No I'm perfectly serious, I am totally pissed off working here, and the powers that be of THF won't allow me to do what I'm good at, which is looking after banquets; I spend all my time in the office or at meetings with the Forte high-ups at Café Royal. If you are starting a City operation, I would love to run it."

Having agreed that John Coomb wanted to join us, and we certainly wanted him, I telephoned Bill Graham, our newly installed chairman, in Majorca, where he was wintering, and told him over a very crackly line that I intended to bring John in as a director and I would like to hear his view.

Bill, still with the powerful Geordie accent that thirty years in South Wales had embellished and modified but not erased, boomed, "Coomb - John Coomb? The last time I saw him was when I owned the Beaufort Hotel at Tintern and he threw ma head waiter out the window." Somewhat taken aback, I replied, "Bill, I didn't know your head waiter, is that a vote for or against?"

We both agreed that the said occasion was a long time ago and John had since calmed down and, in any case, a large part of his earlier wildness had been due to the frustrations of working for a large organisation, the very reason he was joining me. When, subsequently, I mentioned this conversation to John he agreed that he had indeed thrown the waiter through the window and that the waiter had deserved it – but, in fairness, the window was open.

And so, several months later, because Ring & Brymer insisted he worked out his notice, started a very profitable and enormously enjoyable relationship. His arrival in September 1979 changed not just the nature of Paynes Caterers but helped set in train a process that caused it to change its name.

The names Payne & Gunter worked so well together that, during the nineteen-eighties, while we were growing so rapidly, everyone assumed that it referred to two old Victorian characters who set up business together rather like Michael Marks and Tom Spencer in the eighteen-eighties.

In fact there were two dominating characters but they were separated by more than a hundred years, my father Gus Payne, late of Shepherds Bush Market, and William Gunter, doyen of society caterers, who held the Royal Warrant for George IV, William IV and Queen Victoria and was the author of the delightfully eccentric "Gunter's Confectioner's Oracle". The story of how the two names came together is, as usual, one of chance and mischance.

Changing the company name was the inspiration of a sales manager we had for a year or so in the early eighties. He might as well have his moment of glory now for he will not appear again, being neither use nor ornament, although he certainly considered himself both. I am however grateful for the inspired thought he had of renaming the company.

As he pointed out, our existing name was very clumsy, the word 'caterers' undersold what we did, had down-market connotations, and the name 'Payne & Gunter' rolled off the tongue so well and gave us gravitas like our much bigger competitors, Ring & Brymer with their history of "catering through thirteen reigns" and their great rivals, Letheby & Christopher, who were primarily racecourse caterers.

It was inspired and almost at once everybody thought that was what we had always been called. I have to admit that it took some chutzpah on his part to convince his boss that he should change the name of the family company - after all I was the third generation Payne in the business.

The actual involvement with Gunters had started a couple of years earlier, towards the end of 1979, just a few months after John had joined me, with a stroke of serendipity when I heard, through the industry grapevine, that Tom Kingston was trying to sell his family business, the Mayfair Catering Group. This was a disparate group of three companies that had come together over many years, having all started in Regency times.

The companies within the group up for sale were Mayfair Catering Company, (a very high-quality equipment hire company), Kingston Miller & Co, (a catering company concentrating on the West End of London) and Gunter & Co, (which was for many years the owner of a very famous tea shop but for the last twenty had concentrated on City of London banqueting). Given the fact that with John Coomb's arrival we had just started banqueting in the City, it didn't take me long to make contact with Tom Kingston whose great, great, great, great grandfather had started the whole thing.

The original Mr Kingston was chef at Gunter & Co in the 1820s; he departed and, together with the butcher who had been supplying Gunter, a certain Mr Miller, set up a rival business, Kingston, Miller & Co.

The relationship that sometimes develops in the catering industry between chef and butcher was called "perfidious" by Cesar Ritz, and he should know, having sacked the great Escoffier from the Savoy Hotel, for just such a relationship, as did the Reform Club their celebrated chef Alexis Soyer, 70 years previously.

Chefs and butchers don't often run off with each other, as was the case with Gunters, but frequently do so with their boss's profits, usually by the chef ordering, and the unknowing boss paying for, a lot more meat than the butcher subsequently delivers.

Fifteen years after Kingston Miller & Co started trading, Mr Kingston's son fell out with his father and left the business to set up another company, Mayfair Catering.

These three companies, Gunters, Kingston Miller and Mayfair Catering continued in competitive parallel for the next ninety years until the two branches of the Kingston family came together under one ownership in the 1930s.

Gunters meanwhile continued independently but on a downward spiral after the closure of their famous teashop in the 1950s. Then, in the 1960s, Gunters were sold out of receivership to Leonard Kingston of Mayfair Catering, (Tom's father), who offered one shilling more than Lindsay Ring of Ring & Brymer. To make matters worse, Ring & Brymer had itself been owned by Gunters until just after the war. It does appear that the catering industry is just as incestuous, and quarrelsome, as European royalty or the Mafiosi.

I knew none of this, and the underlying currents of ill will, when I started to negotiate with Tom Kingston. It all took place over the Christmas holidays in 1979, the first meeting being at Tom's flat in Paddington and subsequently at the Mayfair Catering warehouse in Hearn Street in the City.

My first visit to Hearn Street, obviously undertaken while all the staff were on holiday, was a serious shock to me. The warehouse, although built in the early part of the twentieth century, was everything gloomy and dismal that one would think of in connection with the pessimistic side of Dickens's London. Low ceilings, no natural light, stone floors and staircases, and almost troglodyte working conditions; it wouldn't have surprised me to see Bill Sykes and his dog in charge. It is tragic to think that while Hermann Goering hit a great many of Wren's fine churches in the City he missed this awful building.

The reason for the shock was that I had been many times to Mayfair Catering's former premises, from which they had moved during the previous five years; these were in North Row Mayfair, a very proud, large, Regency terrace house with great bay windows and very grand stone steps up to the front door traversing a wide area which allowed light to the semi-basement.

I used to go there to collect equipment we hired for any special occasion for which we might be catering, as Mayfair Catering Company was for many years the hirer to the entire industry of specialist fine silver and linen. They also had the largest stock of the fine, spindly, gilt chairs, whose use was obligatory for every fashion show in the West End, and which were stored in an old stable, around the corner in Farm Street. They were not the biggest hire company, in fact they did less than we did, but they had the best.

We used to hire their chairs and tables on many occasions and also some special place mats and silver for one specific dinner we did each year in Southwark. I vividly remember the first time, when I was probably no more than seventeen, going up those magnificent steps and opening the polished mahogany and glass door and stepping back a century or more.

The entire office was separated into cubby-holes by yet more highly polished mahogany and glass, forming partitions, with characters out of Dickens in each one, undertaking various arcane tasks. The time I spent waiting, by the gleaming hard-wood counter, while a very elderly lady sorted out the paperwork on a seriously old 'sit-up-and-beg' Remington typewriter, watching all the to-ing and fro-ing of these characters, was fascinating to me and never ceased to be so on my many subsequent visits over the following dozen or so years.

Each time I went there, once the paperwork was done, I had to take it either upstairs to collect the packages of linen, starched and ironed and beautifully wrapped in crisp blue-and-white-striped paper and the boxes of burnished silverware, or to drive round to the Farm Street stables to collect the gilt chairs. Sometimes I was sent to Eccleston Place in Victoria where Kingston Miller were based, to collect some specialist equipment that was stored there; at that time I had no real idea what Kingston Miller was because as caterers we moved in very, very different circles.

During that time, the early sixties, we at Paynes Caterers were very much looked upon as the 'poor relations' and certainly way down the perceived pecking order of caterers, limiting ourselves largely to sporting events such as golf tournaments, while Kingston, Miller looked after what was left of the 'nobility' and 'diplomatic London'. The following few years' social upheaval decimated their market, and the innovation, firstly my father's and subsequently mine, dramatically increased our market and indeed our share of that market and in the process totally changed the relative positions of the two companies.

Because of the change in my role at Paynes, I had not been to collect anything from Mayfair for several years but it was my memories of the elegance and old-world charm of these buildings, and the people who occupied them not more than a decade and a half before, which made the sight of Hearn Street so shocking to me and made me realise just how low the company had sunk - terribly sad.

Tom Kingston, a slight, rather studious and introverted man, might have been the seventh generation of his family since the company started, but he was in the business due to noblesse oblige, not because his genes had dictated that he should be a caterer. He was by nature a computer man but had to take over as his elderly father, Leonard, who I remembered as a very dominating character, in my own father's mould but without the humour, had, according to Tom, totally lost his grip of the company and was being taken advantage of by his trusted staff.

All three companies had drifted downwards over the past few years; Tom was quite a mild man and was obviously dominated by some of the workforce, who had been allowed by Leonard to get away with more and more over a number of years. Tom certainly did not trust some of his workers and, understandably, would not tell us which ones. He also had not the vaguest idea of how the firm had been run for the past ten years, so it would have been very risky for us to take over the company, with all its liabilities, and what appeared to be a workforce of dubious benefit, when all we wanted was the equipment and the right to use the companies' names and to continue their trade. I therefore offered to buy the equipment, paying for it over five years, and to reward Mayfair Catering Group for the use of the trading names with a royalty payment for any work that could be proved to have come from their introduction over the next three years.

Tom Kingston was interested but he also had a property to get rid of. Our great rivals, Ring & Brymer, were outgrowing their City of London base, and the Hearn Street warehouse, together with all its equipment and the ability to get rid of a City competitor, made Mayfair Catering Group in its entirety a very attractive morsel for Ring's parent company Trust House Forte. That certainly was what Ron Zanre, the THF boss for the West End and City, thought.

However, this was Christmas week, Rocco Forte and his father were at Sandy Lane in Barbados and none of Ron's bosses was prepared, either to make the decision themselves, or to disturb the family so that they could make it. This, coupled with the antipathy between the Kingston family and the Rings, meant that what could, and indeed should, have been bought by the mighty Trust House Forte out of their petty cash, slipped out of their reach and fell into my hands.

It was agreed that we would sign the contracts on Thursday 10 January 1980 and take ownership at the beginning of February. That would give us a couple of weeks to inform their customers that we were going to be handling the Gunter and Kingston Miller catering functions, and the Mayfair Catering hire orders from that date.

The day before the signing was due I discussed the situation with my team and we agreed that, since Tom Kingston was unsure of some of his staff, it would be quite ridiculous for us to allow them free rein over all the wonderful silverware for the two weeks before we took over – the odds were that the cupboard would be bare by 1st February.

I met up with Tom to sign the agreements but told him that, since he didn't think too much of his people, I certainly had no reason to, and a little subterfuge was called for. I said that I would only sign if he would agree to tell all the staff at the end of the day on Friday11th, informing them that we would be moving in at the end of the month. We assumed, and it proved to be right, that everyone would leave, and go to the pub to discuss the matter; any who might have had larcenous intent would be happy that they had a good couple of weeks to take full advantage after they returned on the Monday.

We, of course, were ready, like the Sweeney, waiting in vans and cars round the corner, while Tom gave his talk, until the premises were locked up for the weekend. When we had seen everyone disappear in the direction of the pub, we swooped in and worked non-stop over the Saturday and Sunday, packing up all the equipment, except that required for orders and functions for the next two weeks, and removing it to our warehouse at Chiswick.

We will never know whether such tactics were in fact necessary but we did get all the equipment. There were however two errors made that weekend, one of commission and one of omission, both of which were to have considerable effect.

The first was when Tom Kingston asked me if I wanted to take the furniture, which hadn't been part of the original deal, and without examining it carefully, I said that we would, knowing how busy the hire market in these things was. It was a decision I seriously regretted.

The other, which I never properly found out about for several years, and certainly didn't think about at the time, since I had no idea of their existence, concerned the leather-bound ledgers of Gunter & Co, which went missing. These were, in effect, a hundred-and-fifty-year history of Social London, and included every transaction with the nobility over that period, hand-written in them at the time. With hindsight they were potentially worth more than everything else we acquired.

The decision over the furniture, which besides anything else caused chaos at our warehouse, already full with all the other equipment and which meant that we spent a great deal of time and money moving chairs and tables in and out of rented warehouse space, came to haunt us quite quickly. The famous elegant gilt chairs which had graced the salons of Mayfair for half a century, had suffered from neglect in the same way as the whole company and as well as some of the spindles being broken or missing, all the gold paint was flaking off.

Before the chairs were brought to our warehouse at British Grove they were to be taken from Hearn Street to be used at a major fashion show at Earls Court - 1500 of them. Our van delivered them and our boys set them up around the beautifully decorated red-carpeted catwalk, before any of the organisers had got there, as was normal.

The first I knew of a problem was when I received a phone call from what I can only describe as a screaming queen, in an advanced state of hysteria,

"Bring your bloody van here and remove these fucking chairs – now!"

Since he was obviously in no mood for a debate, I got our vans on the road and went with them to the exhibition hall and, after we had removed the offending furniture, I could see the justice of his complaint as the floor was completely covered in flakes of gold paint – quite a pretty sight but not really the effect he was seeking.

It took me a year to get rid of those 'fucking chairs' –

a year of expensively ferrying them around West London.

The weekend of the clear out from Hearn Street was, for some strange reason, the first time I met Bernard Clements, Gunter's General Manager, one of the few employees Tom Kingston trusted, and someone who was vital if we were going to get any of the Gunter catering business to come to us. He was a lovely man, one of nature's gentlemen and was to turn out to be a perfect foil to John Coomb, whom he was to join in running our City of London business.

Bernard too had been catering for banquets in the City and West End for more than thirty years, the first twenty with Kingston Miller and the rest with Gunters. He quite naturally had his fans and his detractors, although, while we discovered that Gunter's reputation had become quite tarnished over the last few years, I never heard a bad word about Bernard personally. He and John were totally complementary, but whereas Bernard was very smart dressed in his City uniform of black jacket and waistcoat and striped trousers, John, dressed the same way, was very elegant. This was partly because John was extremely tall and quite slender while Bernard was short and portly. They became more and more like Laurel and Hardy.

Bernard had a very lively sense of humour, although it took a while before the guard, which had been applied during the past few years of working against the odds, was allowed to slip and the real Bernard showed through. He was to find the atmosphere at Paynes initially very challenging but ultimately extremely rewarding, when he discovered that as a company we fervently believed that being professional didn't mean that we couldn't have fun doing our job - quite the reverse, we expected to.

That first evening, after the 'cops and robbers' activity outside, Bernard and I sat at the Victorian roll-top desks, brought over from the elegant surroundings of North Row and looking very out of place in the dingy offices at Hearn Street, while all around us the Paynes staff were packing up the Mayfair Catering equipment. Bernard had always very much respected John Coomb while competing with him over the past thirty years. He was looking forward to working with him and it was a bonus when he discovered that his new managing director was also a dyed-in-the-wool, hands-on caterer. I wanted Bernard's view of what I was buying and all the gossip and background to the staff. Since we were not buying the company, we had total freedom to choose which of their staff we offered jobs to and my first question to Bernard was, "Who do you want to carry on working with, or to put it another way, who would you be glad to see the back of?"

Bernard's response was immediate, "Do you mean I can get rid of all the people who have given me a hard time?"

"Of course you can"

"In that case there is only one I really desperately want to come with me - Mrs Marchant - it might also be worth having Young Mr House, though I don't think he'll come."

"Who on earth is Young Mr House and indeed why is he young?"

"Well to differentiate him from Old Mr House, of course."

It turned out that Messrs House, Young and Old, were two brothers in their late fifties who were catering managers for Kingston Miller. It says everything about the way the Mayfair Group worked that Bernard had worked alongside them at Eccleston Place, or North Row and subsequently at Hearn Street for the best part of thirty years and he still called them Young Mr House or Old Mr House and he remained Mr Clements to them. As I said, totally dickensian!

Neither Young nor Old wanted to join us, so it turned out that Bernard Clements and Mrs Marchant, as well as a trainee manager, who worked for Gunters, were the only ones to transfer to Paynes Caterers at Chiswick.

Bernard was absolutely right Mrs Marchant was gorgeous. She was the lady who was in charge of the linen, (who, unbeknown to me, had packed all those blue and white parcels for me over the last twenty years); she was at that time in her late seventies, still sprightly and diligent, and had worked for Mayfair Catering for well over sixty years.

Naturally, given her role, she rejoiced in the name of Blanche, and I was to spend many coffee or tea breaks sitting entranced in her newly installed linen room at British Grove, listening to her wonderful stories of the London Social Season in the nineteen twenties and thirties. She described how she used to arrive at the West-End town-houses of the nobility with her iron so that the Mayfair linen could be ironed in place before a debutante's coming-out ball – absolute magic - and she was still very proud of the wonderful linen she looked after.

Once we got it all back to British Grove we were able to see what exactly we had bought and it was like Aladdin's Cave. It was probably the finest collection of catering silver plate in the country, most of it dated way back before the war, and probably had not been used since; it was from a different, and more gracious, age and I could see it would enable us to give our catering an élan that would really set us apart. John Coomb in particular was in seventh heaven, dreaming of occasions where he could show it all off to best advantage and some stunning ones were to follow during the next few months. Whether these came about because of the equipment itself or because of our greatly increased self-confidence is a moot point, probably a mixture of the two.

Some items didn't lend themselves to being used often, for example in Blanche's emporium there were several embroidered green silk display tablecloths 11 yards long - not many venues had dining rooms with sufficient grandeur to cope with that.

Amongst the boxes and boxes of wonderful silverware we discovered four very large, very heavy trunks, which turned out to contain four silver-gilt candelabras, two 12-branch and two 16-branch, which stood six foot high and were five foot across at the branches. Not many tables could take those. They were quite magnificent in a very, very over the top way; they looked like something out of one of Ludwig of Bavaria's wet dreams. These were the only four items we had valued and the man from Christie's put a price on them alone of sixty percent of what I was to pay for the whole company.

We were to discover slowly from conversations with Blanche and Bernard precisely how this cornucopia of equipment had been used when the companies were in their prime in the nineteenth and first half of the twentieth century. But three pieces of paper that Bernard brought into the office on his first morning jump-started the discovery process.

The first was a letter he had received just before Christmas from the Clerk of the Worshipful Company of Goldsmiths telling him that Gunters, who had been, for 20 years, one of the two caterers on their list, were being removed.

Since, on the basis of the figures given to me before the deal, activity at Goldsmiths Hall was approximately 35% of the group's entire catering turnover, and to get on the very prestigious and profitable Goldsmith's list was one of the main attractions of Gunters, this, quite reasonably, could be said to have been unexpected. It was also somewhat of a surprise that this letter had not been mentioned to me by Tom Kingston, before I signed – must have slipped his mind.

What made it worse was that, given the arcane way the City Livery Companies work, by taking over Gunters, we were effectively barring ourselves from going to Goldsmith's Hall in the foreseeable future; it was in fact going to be a full ten years before we undertook another function for the Goldsmiths Company. Had we not taken Gunter over, we probably would have been number one contender to replace them and enjoyed their business during that decade – oh, the ironies of life.

Bernard hadn't mentioned it over the weekend because he, quite reasonably, assumed that I had been told all about this as part of the negotiations, although it now started to make sense why I had not been introduced to Bernard before the deal had been signed. He said that they had had more than a year of problems stemming mainly from the fact that the equipment-hire tail was wagging the catering dog, and Bernard had spent many unhappy hours waiting at Livery Halls for his delivery of a lunch or a dinner, because the foreman, who worked for Mayfair Catering, decided that his hire orders took precedence over catering functions.

As can be expected I was less than happy at this moment and the fact that Tom Kingston's payments would reflect the shortfall in the business caused by the loss of Goldsmiths Hall did nothing to assuage my anger.

Luckily there were the other two pieces of paper.

The first was a very formal document that turned out to be the Royal Warrant given to Gunter in 1855 by the Emperor Napoleon – seriously fascinating. The second was an invoice from Negri and Gunter Confectioners, at the Pot and Pineapple, Berkeley Square, to the Lord Peterborough, for One Pound Sixteen Shillings, and dated 1786.

John Coomb and I looked at each other and toasted our good luck with some celebratory Champagne as we realised how much fun we were going to have with all this.

10

OUT OF MAYFAIR

Having done the 'moonlight flit' with the hoard of Mayfair Catering Company's wonderful equipment, I now allowed myself a brief moment to savour the unpacking of it and the revelations this brought about. I certainly needed Blanche and Bernard to tell me about the more esoteric items, although quite a few of them even they had never seen in use.

The collection seemed to embrace everything from Victorian wicker picnic baskets to four-tier cake-stands, from silver asparagus-tongs and grape-scissors to the most amazing collection of copper pots - all very tarnished but soon to be beautifully burnished. There were also several hundred covers of a unique pattern of silver plate cutlery called Albany with the handles spreading out towards the end in a classical fan, made especially for Mayfair Catering. It had been purchased just after the war but most of it was still in the original boxes from the cutlers in Sheffield – never having been opened.

All the silverware was stamped with the Mayfair

Catering Company's initials, MCC, which occasioned the Master of the Worshipful Company of Salters to be somewhat nonplussed and embarrassed when the Duke of Edinburgh, their guest at a dinner for which we were catering, asked him if the Salters, one of the most illustrious and wealthy City Livery Companies, had nicked the cutlery from Lords Cricket Ground.

My first job, having bought all this equipment was, of course, to make it work for us, more particularly to use the unique nature and elegance of it to enhance our catering product and enable us to get more business and, just as importantly, charge a higher price. Immediately we had some beautiful photographs taken of the more exceptional items and included these in our quotations. This certainly differentiated us from our competitors. Next, John and I had to find some high profile functions and locations where we could show all our wares off to full advantage.

The first one came about by a wonderful piece of serendipity. Sir Tom Collett, for thirty years a very good friend of John Coomb's, as John had looked after him throughout his City career, most especially serving him at the Mansion House when he had been Lord Mayor, wanted John to do a party for his golden wedding.

Lady Collett – bless her – decided that she wanted to have a tea party and so all the Mayfair finery was able to be displayed, silver tea urns and kettles, fine china tea plates with a cobalt blue surround and gilt edges - made by Thomas Goode, a neighbour of Mayfair Catering, round the corner in South Audley Street – tiny flat ice cream spoons, those four-tier cake-stands and even Blanche's famous green silk buffet cloths. It was a wonderful display - made all the better for the fact that Tom Collett's standing in the City meant that everyone who mattered was going to be there to experience it – we should have paid Tom for the honour.

The party was to take place in Fishmongers Hall, which I had never previously visited; Gunters were already on the list of caterers, while Paynes Caterers were not. Fishmongers Hall is one of the grandest in the City; it is the massive building with classical portico immediately on the upstream side of London Bridge on the City side of the river. The current hall replaced the baroque one which had to be demolished in the 1830s to make room for the rebuilding of the bridge – the bridge subsequently sold to Arizona when it was replaced itself in the 1970's – and the fishmongers of the Regency era certainly did themselves proud.

I'm not sure what I expected from the Fishmongers Hall when I first went there. I probably echoed the thoughts of Annigoni who, when commissioned by the Worshipful Company of Fishmongers to paint the famous portrait of the Queen, was initially very sniffy about whether he wanted to be associated with a group of fishmongers. However, when he visited the hall and discovered that it was quite as magnificent as anything Venice or Rome could offer, he decided that their money was quite acceptable and the portrait now has pride of place in the Court Drawing Room.

Bernard Clements, who was a regular visitor to the hall with Gunters, showed me round and let me into the first of the many little secrets of the City that were to delight me over the next few years. Alongside the very grand staircase there is a magnificent stained glass window that replaced one destroyed by enemy action during the war.

Bernard made me look very, very closely at a piece of the lead work and in a certain position I could just make out some tiny words engraved in the soft metal:

"I, Roy Merrivale, aged 18 did all the work but received no credit"

What an insight! I could instantly imagine the apprentice glazier in the late 1940's getting more and more aggrieved by the fact that he was doing it all - only for his well-paid boss to take the glory. Since it is still there I presume that he was never caught otherwise the boss would have removed it – and him.

The party was a delightful occasion and a fantastic shop window, particularly since Sir Tom singled out John Coomb for praise in his speech. So, six months after John had joined me, and less than three months after the purchase of Mayfair and Gunters, we were truly visible in the City; what we now needed was to be properly anointed by being accepted by the Corporation of the City of London to undertake one of their State occasions.

The Corporation organises three or four State Banquets each year. Typically during a visit by a foreign head of state, a banquet is held in their honour at Guildhall when they are given the 'Freedom of the City'. This usually follows an opening banquet at Buckingham Palace and a return engagement at the visitor's embassy, if the facilities permit or, if not, at Claridges or the Savoy. There are also banquets and receptions to celebrate significant occasions, such as Coronations, Royal Weddings, the end of the Falkland War and so on.

Guildhall, where these take place, hold a list of a dozen caterers who are allowed to work in the building – which at the time included Kingston Miller and Gunters – but there are, at any time, only two or three who ever undertake the prestigious state events, rather than the hired-hall occasions that happen virtually every day.

Since Sir Lindsay Ring, John's erstwhile boss, had been Lord Mayor three years prior, his own catering company, Ring & Brymer, had picked up all the prize functions for a long time, and while Lindsay was still a considerable force, there was a feeling that it was time for a change. Our moment was about to come.

John Coomb, at this juncture, had done ninety-two such occasions – he was to notch up a hundred and twelve by the time he retired (a figure which will never be equalled) – and he was extremely popular. It had been decided by the City bigwigs that Paynes Caterers should be tried out on a reception (a very elaborate buffet) before being allowed to undertake a full-blown banquet. It just so happened that there was to be one – and it was going to be a cracker – that summer to celebrate the fortieth anniversary of the Battle of Britain, a reception and supper for fifteen hundred guests at Guildhall with the Queen Mother as guest of honour. What a way to start!

We had to go through the whole process of appearing before the reception committee, which I found fascinating and quite intimidating. Unbeknown to me, it was in fact a charade because the decision had already been made.

The outcome was that we were going to be undertaking the catering for this outrageously over-the-top function as our initiation into the City big-time. The trouble was - as always 'Sod's Law' works with function caterers – it was to be in four weeks time, in July, plum in the middle of our frenetic outdoor events season. We were still primarily an events company and seventy-five percent of our business took place between May and September, mostly on golf courses and more recently on the banks of the Thames at Henley.

This year was no exception. Between the last week in June, when it was confirmed we had the Guildhall reception and Monday the 22nd July when it was to take place, we had a frantic diary. Starting with Henley Royal Regatta where, five years earlier, we had introduced hospitality pavilions, and which was at that time by far our most profitable contract, followed by a professional golf tournament, fourteen other functions and finally the Open Championship. This, the proudest and oldest name in golf and still, in 1980, by a considerable degree our biggest event, was to take place at Muirfield on the Firth of Forth and due to finish on Sunday the 21st July.

It was a pretty scary timetable. John was going to have to look after Guildhall on his own – I would only be able to get to the function at the last minute, it would be no hardship for him on his ninety-third state occasion but hell for me four hundred miles away while all the preparation was done.

To make matters worse, before we knew about the Battle of Britain reception, I had decided to make the Open Championship this year into a real showcase. As well as serving some sixteen thousand lunches to our clients' guests during the Championship week, we would, as usual, be entertaining up to a hundred each day in our own pavilion. The difference this time was that I had planned to fly in new customers from the City of London to experience the scope of our activity.

To make it even better, and more onerous, I had rented a huge house in North Berwick to accommodate our visitors, where we would also be able to have large dinner parties so that our new customers and the golf world could meet in a social setting.

The house, Glenconnor, was an Edwardian Scottish-Baronial pile with red-tiled turrets and gables and granite facings with redbrick highlights. It had more than twenty bedrooms, several reception rooms including a music room, and a dining room that could comfortably seat thirty. The

Glenconnor family had built it so that they could spend the shooting season with David, Prince of Wales when, in the nineteen twenties, he moved up to Greywalls at Gullane, next door to Muirfield. This happened every summer after the Glorious Goodwood Festival had finished, which indicated that the 'London Season' was at an end.

It must have seen some very stylish parties, but was nowadays used by the family for summer holidays of a much more modest nature and seemed very sad and dismal when we first saw it on a dreary April day, shrouded in morning mist. However, I knew even then just how wonderful it was going to be as I could envisage what it could be like.

It didn't look much better when we first arrived in July but once I had set our team to work, filling it with our Gunter goodies, candelabras, silver vases, wonderful linen and fine crystal and most especially flooding the entire place with masses and masses of flowers, Glenconnor seemed to come to life – it was as if the house was saying, 'Yes, that's what I'm supposed to be for - entertaining'.

It was seriously hard work but enormous fun, made possible only by my wife Judy. At that time she worked full-time for the company in charge of the petty cash, which in a business like ours was far from petty. I gave her the nickname 'Payne-the-purse', and she more than earned it.

As I mentioned before Judy was a natural hostess and she thrived on the rôle of châtelaine at Glenconnor; although how she did it as well as looking after the cash at the Open I have no idea. As soon as we finished the lunch session at Muirfield we were off to North Berwick to greet the new arrivals from London and prepare for the thirty guests for dinner. And we did this for six days.

Champagne in the music room, with a pianist playing Bach's 'Well-tempered Clavier', followed by the dinner itself prepared by our top chef, served on the mahogany dining table that had almost certainly welcomed the Prince of Wales and Mrs Simpson. We, and our guests, certainly had a lot of fun, with TV celebrities, all long-standing friends, such as Peter Alliss and Harry Carpenter - the sound and the face of the BBC team – and international golfers Graham Marsh and Tom Watson, Hale Irwin and Lee Trevino mingling with the guys from the City who, as 'golf nuts', thought they had died and gone to heaven. Very good for business too.

One of the people I invited was the local director of my bank, as it always makes sense to show them what you're doing with their money and indeed how you earn yours. Although he had given us no help with the mortgage for British Grove, he still provided our business facilities and overdraft. He flew up at our expense with a friend, stayed overnight in Glenconnor, as my guest, and was mightily impressed with the evening and his fellow guests. The next day we took him to Muirfield and, after following the golf all morning, he came into our pavilion for lunch.

One of the things I wanted to get across to him was the scale and scope of the company. There had always been a tendency among customers and suppliers to think that, because the company had my name it was effectively only about me – that I was its only asset, and I knew that the man from Barclays thought the same. I certainly had no false modesty about my importance in the scheme of things at Paynes Caterers but I knew very well I was not a one-man-band - after all there was a little affair in the City of London being prepared as we were enjoying ourselves on the East Lothian coast.

The next day when he arrived at our pavilion it was, as per usual, the standard organised chaos with guests coming in and out, following matches, wanting lunch, having drinks; other regulars just calling in to say hello. The Open Championship is the Mecca of golf and we were an integral part of the golf tour, catering as we did that year for thirteen tournaments, so all variety of visitor just popped in; there was a pleasant tumult of, by and large, lovely people, together with a few free-loaders.

I decided that I ought to sit down with the Barclays man and his colleague and so started to gather together a full table of guests. It just happened that Sir Matt Busby and a friend of his, Harold Riley, (the Mancunian artist and pupil of Lowry who, twenty years later, was to give the address at Matt's memorial service), came in at that time with Peter and Jackie Alliss, so the four of them made up our table. Wonder of wonders, towards the end of the meal Matt came out with a question that I could have written myself, given the audience at the table, "Alan, how on earth are you able to sit down and have lunch with the likes of us while you're feeding five thousand people here today?"

My answer, ostensibly to Matt but really addressing the other side of the table was, "It really is the same as you Matt, you didn't play for United, you told the boys what to do and sent them out to play – half time you gave them a pat on the back or a bollocking and sent them out again. I'm the same. There's no way I can serve five thousand lunches myself so I have to rely on a very efficient team which, luckily, I can do"

The message got across - particularly when I told him what John and the rest of the team were preparing for in London in my absence - and at the end of the year this whole episode was going to pay serious dividends.

Muirfield, or to give it its proper name, The Honourable Company of Edinburgh Golfers, is a wonderfully eccentric place, full of characters out of a Somerset Maugham novel. The members have, historically, chosen to be organised by a series of bullies as club secretary, employed to keep visitors at bay with a withering line of unbelievable rudeness.

The Secretary in control at the time of the 1980 Championships was Captain Paddy Hanmer who, reputedly, would not allow Keith Mackenzie, the Secretary of the Royal & Ancient Golf Club of St Andrews, and as such the organiser of the Open, to play a round of golf at Muirfield following a preliminary meeting there with the BBC. It seems Paddy looked out of his window, saw a couple of members starting their round and so told the most important man in golf that the course was too busy and that the Secretary of the R&A and the BBC producers would not be able to play.

My previous experience of Muirfield was seven years before when we were preparing to cater for the Ryder Cup Matches and I had to go there for a meeting with the PGA. It was the first of the many reconnaissance visits that all events require. I had come to Scotland by sleeper into Edinburgh. I arrived on a freezing January morning to a less-than-welcoming Waverley Station, picked up my hire car and thought that, although I was early, I might as well go straight to Muirfield where I was sure to be able to have a warm-up and a breakfast. Wrong and wrong.

I got to Muirfield at about a quarter to eight, and the meeting was not scheduled until ten so I ventured up to the front door and knocked. The groaning door eventually opened, as in a Hammer Horror movie, and a very dour Scottish functionary, after some hesitation, bade me enter, and took me through to the men's lounge, which was even colder than outside. There were no fires and certainly no heating on but he told me to wait here. It was only a couple of hours after all!

I sat down on a very old, very hard armchair and the

cold of the worn leather went straight through my clothes and literally chilled me to the bone. I hoped that he might offer me a tea, or at the very least light a fire in the enormous baronial room with its massive windows looking out onto a very bleak golf course, but instead yon man proceeded with what appeared to be the necessary morning preparations for a noble Scottish golf club.

Firstly, he brought in the whisky trolley, which had perhaps forty different malts on it, together with 'the book' (this was a little note book with a pencil both attached to the trolley with string, so members could jot down the drinks as they had them).

Having duly set this up he then disappeared only to return with the liqueur trolley, similar to the whisky one but with every possible version of what the French call 'digestifs', including, of course, several bottles of Kummel, the aniseed liqueur without which no Scottish golfer can possibly face the weather. This trolley also had its own 'book'. Both trolleys and books were in place by 8.15 of an icy January morning.

He still had not offered me a tea or coffee or indeed turned on the heating but he did now address me with a question, "Would you be wanting a Scotsman?"

Now, this was not something I had ever been asked before but I thought it would at least keep me warm. It was then that I realised that he was referring to the local newspaper - altogether a strange experience.

Anyway, forward again to 1980, the week of the Open at Muirfield went like a dream. Seriously hard work, quite exhausting, but very successful from the point of view of the business for which we were being paid that week and also for the seeds we had sown both in our pavilion and at Glenconnor.

Monday morning, after a final night's dinner at the house, saw us at Edinburgh Airport on the way to London and the reception at Guildhall that evening. We shared the journey with Tom Watson, who had won the Open the previous day and his wife, Linda. They had dined with us at Glenconnor on the first night we were there and so it was a very jolly journey, with Tom proudly carrying the famous claret jug – it's amazing that even though he had already won the Open a couple of times before, he was still like a big kid – I suppose even the stars never take this trophy for granted. During the flight my mind was starting to race, hoping that John Coomb was well advanced in his preparations for the Queen Mum – I knew he was, but I still worried.

So a quick trip from Heathrow home, to drop Judy and our luggage, and get my dinner-suit for the evening and then to our office at Chiswick as I had a few hours work to catch up on before going up to the City.

What I had neglected in the list of clashes that came about when we were given the contract by the Corporation was that on the Tuesday – tomorrow - I was due to be flying off with my family for a month's holiday in Canada, so I had a lot to deal with.

Late afternoon I drove up to the Guildhall and parked in the space allocated to us in the car park underneath. When I went up the back stairs from the caterers' loading bay I felt the inevitable tingle of anticipation I had always felt before any major event – it was probably the reason I did this silly job!

Of course, other than in the planning, I had had no part in the preparations for this reception or indeed in its execution and, since it was the first major function that we had done at Guildhall I was seriously excited and more than a little nervous and the sight that greeted me filled me with wonder. I was to see it have the same effect on many others over the next two decades.

Guildhall itself with white clothed buffet-tables around the walls looked spectacular and our splendid display, with the Mayfair silver twinkling dutifully, was quite overwhelming. The Pikemen of the Honourable Artillery Company in their fifteenth century uniforms with steel helmets and breast-plates set off the whole scene, standing guard in Guildhall and acting as muster points for the guests; set against the chequer-board stone floor, the whole thing looked like a giant multi-coloured chess set.

In the crypt we had a display of thirty different cheeses and a cornucopia of fresh fruit and shelled nuts; in the Old Library there were two buffets, each fifty feet long, adorned with our massive 12 and 16 branch silver-gilt candelabras, holding 56 two-foot-tall candles – at last we had found somewhere large enough to use them – with a medieval display of dressed meats including boars' heads and roast geese. I know I had seen it written down in our quotation and had helped John describe it to the committee but, actually seeing it all on the evening, I was very impressed, and that didn't happen often.

The first guest I saw was Sir Douglas Bader, the legless Second World-War fighter pilot, not surprising at a reception to celebrate Churchill's 'heroic few' since he was one of the most famous, having had a film made of his life. He had been a great friend of my father Gus through their mutual friendship with Henry Longhurst, doyen of golf writers, commentators, raconteurs and drinkers. For some reason Douglas looked very shaken so I took him into our wine-bin next to Guildhall and poured him a drink.

He told me that he and Henry Longhurst, who had

died a year or so earlier, had agreed in their cups one night, many years before, in the days when they both still played golf, that whichever one of them died first would get a message back to the other to let them know if the game is played on the other side and if the grass is indeed greener.

He then said in a voice that had less than his usual resonance, "I had completely forgotten about the conversation until this evening when a lady came up to me and said, 'Is your name Douglas?' I said it was and she said, 'I am a medium and I have a message for you from someone called Henry – he says the grass is greener'"

Douglas Bader, a man not easily shaken, was clearly spooked by this, and so was I.

Soon after this I left Guildhall in John Coomb's capable hands as I had several hours' work to do at British Grove before my holiday, so stopping only to buy a hamburger, I drove back to Chiswick. About three o'clock in the morning I had finally tied everything up and made my way home.

Judy was used to me coming in at all hours and didn't usually wake up but she did this time and wanted a detailed description of the whole evening including a blow-by-blow account of the buffets and all the food on them. She then said, "So what did you have to eat?" "A Big Mac." I made sure the press didn't get hold of that story.

The next morning we flew to Montreal, where we were staying with some friends, and they took us, very jet-lagged, to a party at a friend's house almost as soon as we landed. The hosts also had some houseguests from the UK and naturally we started to chat and Judy mentioned that we had been in North Berwick until the previous morning.

She could not believe it when the response was, "Oh! We have a family house up there, which our cousin let during the Open – it's called Glenconnor".

One hell of a coincidence, especially since they had expected to leave Montreal the day before we arrived.

A week later we drove to Toronto where we stayed with Judy's great friend, Mo Dixon, who had moved out there a couple of years before, and went to our first baseball game at the Blue Jays' Stadium, which was a significant visit not only because there happened to be an earthquake during the game but because I saw in operation at the ground a system for pouring and sealing large quantities of drinks in wax paper cups made by the deliciously-named Sweetheart Cup Manufacturing Company of Pittsburgh.

I had no idea at the time but that was going to be by far the most valuable discovery I ever made in my entire business career and, quite literally, was going to transform the company within six months.

11

Sweetheart Deal

It was fortunate that I felt invigorated by my holiday in Canada, because I arrived back to chaos – organised chaos, of course. Certainly, what I found were the problems of success, but problems nonetheless. I returned to a very full in-tray of thank-you letters from people we had entertained at Muirfield both within the pavilion and at Glenconnor. The whole exercise was definitely going to pay dividends. While I was away, as well as looking after a small golf tournament, my events team had started preparations for the final flourish of events in September and October.

The Battle of Britain reception at Guildhall the night before I left had been a resounding success, which had made Paynes Caterers' City of London operation, led by John Coomb, a fully accepted caterer for all parts of the City hierarchy. The Mayfair equipment had been taken to the heart of John's City clients and the end of his first year with us he was starting to bear fruit. John was always one of the greatest caterers in the land, immensely popular and trusted but, in his previous role, was not allowed to do what he was born to do, namely serve people with magnificent, brilliantly-orchestrated banquets. For the last ten years he had been expected, in his exalted position, to sit in an office or chair meetings - both anathema to him.

Now, given his head, in his mid-fifties, with the company's full support and, of course the stunning armoury of Mayfair and Gunter equipment, he was taking the other caterers in the City, including his old firm, Ring & Brymer, to the cleaners – and enjoying it immensely. But the City's season did not start again until the end of September and in the meantime John was already involved in setting up the massive British Aerospace operation at Farnborough Air Show the first week of September.

My father Gus had been catering at Farnborough since 1948 and he always considered that the money he earned there in September was his "winter keep". He was most upset when in the 1960s it changed to a biennial event but thus it had remained. It therefore added a great deal of lustre, not to mention pressure, to our events season when Farnborough came around.

The biggest contract at Farnborough was for British Aerospace and Paynes Caterers, although we looked after fifteen or so companies at the show had always been considered too small to look after Aerospace, which traditionally had been the province of Ring & Brymer. John Coomb's arrival, together with all the new equipment and the events - including Muirfield – to which we had taken the Aerospace decision-makers, demonstrated to them that we were to be trusted with this very exciting contract. This added to an already full show and a September and October calendar bursting with three golf tournaments and the Motor Show, which meant that I spent the rest of August in detailed planning. I very soon forgot all about my three weeks in Montreal and Toronto and my - as yet unappreciated - visit to the Blue Jays.

Farnborough, with British Aerospace's fifteen hundred lunches a day doubling our numbers for the week, was incredibly hard work, it always was. Firstly you spend the best part of two years before each show competing with the entire outdoor catering industry in order to win each of the hundreds of individual contracts for the entertainment chalets. Then, having won some of these contracts, you are then competing again for all the staff, temporary management, chefs, suppliers, accommodation, preparation sites, access – it really is a true marketplace and the caterers certainly earn their profit that week.

Exhausted but exultant, John Coomb and I had a very amusing dinner with our team at the end of the show. We always had to let off steam after such a week and the glowing reports from all our clients made for one hell of a party, with John retailing stories about the early days of Farnborough when it was run by some of the guys from the RAF who had saved Britain, and indeed civilisation, during World War II. Although we might have been acting at the end of Farnborough as if it was 'the end-of-term', it certainly wasn't. We were going to severely test the twin-track structure of events and City that we had put together when John joined us a year earlier.

We were now going our separate ways - John to the City to start his new banqueting year, which always commences after the Livery Companies and the Corporation come back following their total closure during August; John and Bernard Clements had a very hectic calendar in front of them.

I was going back 'on the road' with my end-of-season flourish of major events. I hardly saw John for several weeks as the events department had a dramatic finale to its season, following Farnborough with the Dunlop Masters, World Match-Play Championship at Wentworth and a new and very different operation at the Motor Show in Birmingham.

While we were certainly great innovators in the world of event hospitality, there is no doubt we didn't achieve it all on our own. I was very fortunate to start branching out and upgrading our product during the last few years of the seventies at the same time as some other complementary revolutionaries were adding to the exciting mix.

Most notable of these was Martin Devetta, a tough, fast-talking, hard-driving, brash individual who took a brilliant design by his engineer father, John, and turned his company, Spandrel Domes, into the standard for the tentage industry. Prior to his arrival on the scene in the mid-seventies, all tents had been ridgepole and guy-rope affairs, not having changed for hundreds of years.

As a personality, Martin was, and is, a one-off, of whom it is possible to believe almost any of the outlandish stories that surround him. A serious golfer and a popinjay of some note, to the extent that he is known at Sunningdale Golf Club – his spiritual home – as 'Cashmere'. The outrageous and delightful Pat Heneghan, organiser of the Irish Open, always referred to Martin as "the man in the Gucci Wellies".

Martin's love of golf was not allowed to be neglected just because of the immense pressures of running a very successful tentage business and this sometimes caused some hair-raising moments. As a member of the Royal & Ancient Golf Club of St Andrews, he liked to play in either their Spring or Autumn Meetings and so one year, because of event timetables, he had to catch a last-minute sleeper from Kings Cross to enable him to play the famous links the following morning.

Martin chided, and doubtless rewarded, the conductor

of the sleeper car to make sure the man woke him in the morning and got him off the train when it made its brief stop at Leuchars Junction, a mile from St Andrews, regardless of how unwilling he, Martin, might be. He then, after a glass of malt whisky, went off to a well-earned sleep, dreaming of the birdies he would find on the Old Course in the morning.

He awoke to hear a severe clanking as the train was slowing down and on lifting up the blind discovered to his horror that they were pulling in to Aberdeen. Not surprisingly, this occasioned Martin to use some of the epithets and curses that kept his workforce so docile, on the hapless conductor, whose response was,

" Ye're certainly a fair one with the invective sir, but nowhere near as good as the man I put off at Leuchars at five o'clock!"

When we first started working together, Martin and I had a lot in common, being the same age, both with the dubious benefit of following powerful fathers, and firmly believing in raising standards – and prices. I was more than a little instrumental in helping him develop his business, as I believed that Martin's new, exciting designs, based on Buckmaster Fuller's geodetic domes, made of aluminium tubes with blue and white panels stretched underneath, were going to revolutionise the tentage trade.

The help I gave was in the face of considerable opposition from Gus, who rigidly believed that the more money the event sponsor spent on tentage, and Martin's were considerably more expensive, the less he would spend on catering. I believed, and was proved correct, that serving our customers in these marvellous new structures would greatly enhance our product and allow us to continually improve and what's more, charge more – dramatically more.

It was because of this belief that I helped Martin persuade Richard Brown, golf boss of Dunlop Sports to adopt Spandrel Domes for the Dunlop Masters in 1975 (to be fair, Richard didn't need much persuading but his bosses did); at that time this was one of the largest events on the tournament circuit.

The vast Colgate-Palmolive Company, who, in the mid-seventies under a golf-mad and hard-drinking chairman, David Foster, had bought their way into the world of golf and were using the sport to promote their image worldwide, also took up Spandrel in a big way. Within a couple of years they had ownership of three tournaments in the UK, all using the combination of Spandrel and Paynes Caterers.

By 1977 Martin's father had designed a new, larger and far better concept, created with our hospitality pavilions complex at the Open Championship in mind. Because I believed they would considerably improve the operation, I exerted whatever pressure I could on the R&A to use them at St Andrews the following year.

That year, 1978, was a nightmare for anyone in a time-critical job – and they didn't come more time-critical than event catering – as the trade unions throughout the country were running amok and both manufacturing and transport were seriously affected. Creating Martin's new tents from scratch that summer was a very risky business and I was severely tested in my belief in Martin's ability to deliver, wondering all the early part of the summer whether I would have any tents in which to serve my sixteen thousand customers at St Andrews.

When we arrived at the golf course, which sits in a bay on the coast of the Kingdom of Fife, we were somewhat dismayed to see that the new structures were still being built. They were extremely complicated and this was very much a learning curve for Martin's team. My team, meanwhile, unable to start our four day set-up, were playing football on the wonderful sands that were to feature in the film Chariots of Fire a couple of years later.

For me it was a little more stressful but I needn't have worried, as Martin, of course, delivered just in time and the structures were quite as spectacular as I had imagined – they took the whole industry up several notches in style. They were made up of massive aluminium struts forming 100-foot spans from which alternating blue and white plastic sheeting was slung in fifteen-foot sections, which could continue to almost any length.

This gave us a vast area without any of the tent poles that all previous tents had suffered from. It enabled us to have two forty-foot deep pavilions separated by a central twenty-foot kitchen. As well as being extremely practical it also looked stunning. It was Gus's last Open Championship and, indeed, his last summer and, in the end, he was glowing in his praise for Martin and his revolutionary new-style tentage.

It was an outstanding success, deservedly so, and in the following couple of years, Martin and Spandrel took over an increasing number of events. The trouble was that these were concentrated in the summer months and the build-time for Martin's tents was very much longer than for the traditional ones. Whereas old-fashioned tentage for a major event would take a maximum of a couple of weeks set-up and a couple of days take-down, Martin's new structures took six to eight weeks for an occasion like the Open and a week or two dismantling. This meant that he required an increasing amount of stock to cover successive events, also knock-on delivery problems became an increasing dilemma, not least for the guy who was going to use the tents – me.

Martin, chubby-cheeked with a full head of wavy hair, always immaculate even when working on site and I, bearded, balding and alternately in scruffy garb while setting up events, or dressed very formally when 'on parade' at functions, became a common duo at most of the summer events.

However, we actually spent more time together during the winter, at site visits and planning trips, than we did during the summer itself. In the season, given the nature of our roles, we tended to leapfrog each other and our meetings usually comprised one or other of us giving the other a bollocking for being late or dealing with some cock-up perpetrated by a member of our teams.

We actually worked well together, trusted each other and there is no doubt that my life was considerably enriched by meeting and working with him. That comment is not, however, without its irony because although our two companies worked magnificently in harness together, the joint ventures and companies we set up together over the years had a very chequered record. In fact twenty years later he was to lead me into by far the most expensive mistake of my life – but that, as they say, is quite another story.

It was due to one of these joint ventures that I found myself, less than two months after my return from my family holiday in Canada, in a muddy field in Warwickshire owned by a farmer called Higginson, wondering whether my team were ever going to be able to set up for, let alone serve, another bunch of several thousand customers. I had quite forgotten the visit to the Toronto Blue Jays and the Sweetheart Cups. However, soon they would come to the forefront of my memory.

In the meantime and also in the mud, Bill Quinn, Martin's no-nonsense-take-no-prisoners site boss, tall, rugged, with the look of Burt Reynolds about him, was wandering around in frustrated rage waiting for equipment to come from those other events for which we were also catering, the Dunlop Masters in Chepstow and the World Matchplay Championship in Surrey. In successive weeks, both suffered from appalling autumn rain that made striking the tents an even longer process, and extremely dispiriting and messy.

The piquancy of the position in which we found ourselves over in Mr Higginson's field was that the several thousand customers who were due to be fed were just as much Martin's as mine. We had formed a joint venture company called Caterdomes – I was the cater, he the domes – or if you like, he was Stanley Matthews and I was Tommy Lawton, only, if he didn't kick them over, I couldn't nod them in.

We started this company to create and sell hospitality facilities so that the motor trade could entertain their customers in great style at the Motor Show, given the extremely limited facilities that the still new National Exhibition Centre offered. We had been very successful at the selling, with a team working on it for over a year, and it was now a question of whether we could create the facilities in time for our clients to entertain the large number of people they expected to visit them at the show.

The day before we were due to open, Martin arrived in his exceedingly large gold Mercedes and put on his "Gucci Wellies" before walking round the site with me, getting increasingly agitated as the gravity of the situation became apparent. A council of war was needed and Bill Quinn was in the hot seat.

"OK, Bill," said his boss, "no bullshit, where exactly are we?"

Bill's response made even the battle-hardened duo of Payne and Devetta blanch,

"We are a thousand man-hours behind the game". When he finally found his voice, Martin exploded, "What the fuck are we going to do?" Quietly, laconically, without even the hint of a smile, Bill said, "I thought I'd get a thousand men in - for an hour"

The explosion of laughter from the two of us had the desired effect of puncturing the near panic that had gripped us and we set our two teams to work with a will. We worked through the night as van-by-van, trailer-by-trailer, the tentage and fittings arrived from St Pierre and Wentworth; given the look of it, the equipment could have come from Passchendaele or the Ypres Salient.

During the evening while we were working our joint socks off, I got a call on my car phone. Martin and I each had one of the first mobile phones in the country, not surprising given our jobs, but this was 1980 – these were firmly anchored to the cars and very rudimentary. Like a war-time walkie-talkie, it was necessary to press a button on the handset to speak and let go to listen and required the people at both ends to know and remember this, because if the person the other end spoke hurriedly when you had stopped, you didn't hear the first few words they said.

Bernard Clements was on the phone from British Grove where he was quietly catching up on his paperwork, the one evening that week when he didn't have a dinner to attend. He had never made a call to a car phone before and had no idea of the necessary protocol - he was also very excited.

He had been given an urgent message from John Coomb to pass on to me and here I was on the end of a very indistinct line, in a field in Warwickshire, very angry, very wet, very exhausted and, hearing only every other word, I was very far from being a happy caterer.

Mustering as much calm as I could, realising that losing my temper was only going to make the situation worse, I managed to get Bernard to stop responding the second I stopped talking by using the wartime R/T rule of saying 'over' as I let go of the button,

"Right Bernard, what is the message? Over."

"John Coomb has phoned - he's in a pub with the Secretary of the RFU. Over."

"Good for him, do I need to know this? Is he misbehaving? Over."

"Well the current caterers at Twickenham are quitting after the Varsity Match and the RFU want us to take over. Over."

"What? Over."

"The current caterers at Twickenham are quitting after the Varsity Match and the RFU want us to take over. Over."

"That's what I thought you said. Was he sober? Over."

"Reasonably. Over"

"When are we to start? Over."

"The Scotland match in February. Over"

"Bloody Hell. Sorry, Over."

He couldn't tell me any more, not even which pub John was in, so I let him get back to his contract sheets detailing all his requirements for the next week's functions. I was left in a state of total turmoil. I carried on all night with the preparations for the Motor Show pavilions while turning over in my mind the possibilities of what, if Bernard had got the message right, might be available for us at Twickenham.

I knew that the year before, just after John joined us, he and I had a meeting with Peter Yarranton, a very much larger-than-life member of the RFU Committee, who had been delegated the task of setting up a hospitality operation for the Home Internationals at Twickenham and wanted to see what we had done at the Open and Henley. Were we interested in setting up a similar operation, which, since it would have been catered by us, would have meant splitting their existing catering contract?

In the event, he decided to hand the whole thing over to the organisation that had started the hospitality facilities at Wimbledon with the RFU's existing caterers, Ring & Brymer, servicing them. The RFU had allocated 400 tickets for each of the matches against Ireland and Wales during the 1979/1980 season. Although only a couple of hundred were sold for Ireland, the full four hundred for the Welsh match were used.

I hadn't heard that it had been a catering disaster (and I certainly would have done) so that was not the reason for this very sudden change. Here I was in a muddy field in Warwickshire, going through the motions of setting up a key event and at a loss to know the cause of this considerable change of heart by our major competitor – to drop a contract mid-season is virtually unheard of.

With all of us having had very little sleep, somehow we managed to get the Motor Show operation up and running by the next morning – it looked fantastic and it needed to be as this was Press Day. We were about to serve more than two hundred of the most pampered and indulged of the land – quite a test after all that had gone before. It proved once again that our true motto like the theatre should have been 'It'll be alright on the night' – only we never get a dress rehearsal.

It wasn't until quite late on that day that I was able to get John Coomb on the phone and hear the whole story about Twickenham. Apparently, according to the RFU, Ring

& Brymer had had considerable difficulties for many years in serving the very large demand for beer, causing their staff to take short cuts in pre-pouring and so falling foul of the weights and measures laws. This culminated in them suffering fines of more than two thousand pounds during the previous season. Sir Charles Forte, as he then was, stated that he was not in business to be fined and that if they couldn't do it correctly they would withdraw. Hence the reason for our sudden opportunity - assuming, of course, that we could overcome the same problem.

John went on to explain that the facilities at Twickenham were very, very ancient and rickety. The electricity available was not sufficient to have electrically-operated measured pumps for serving the beer. Under the 'Safety at Sports Ground Act', which had been brought in after a decade of mounting soccer hooliganism, it was not permitted to serve beer in glasses with measures printed on them, since these were deemed to be potential missiles (even at rugby grounds which had never had any crowd problems). Only paper or plastic containers could be used which, by law, had to be filled in full view of the customer.

Since some seventy-thousand beer-mad rugby fans all wanted a pint at half-time (in ten minutes) and then again at full-time, and each pint took thirty seconds to pour, the ninety pumps in the eleven bars were not really up to it. Had they been asked whether they wanted a fast pint or a full pint, there is no question what the rugger-buggers would have chosen. But the weights and measure legislation was there to protect them from themselves.

The problem was compounded by the fact that the weights and measures man was himself a mad-keen rugger fan from Cornwall, which resulted in the amazing coincidence that Ring & Brymer had a visit from him at almost every match, rather than the cursory annual or biennial visit most other operations received. Of course, if we inherited the contract we would also inherit the problem.

It was necessary for me to remain at the NEC for the first few days of the Motor Show but, while starting the operation, and serving the first of the many thousand anticipated for the two weeks of the show, I was turning over the problem of Twickenham's beer in my head.

A 'Eureka' moment the following night set my brain racing. I awoke in my hotel in the middle of the night having seen in my dreams the Sweetheart Cup Company's machine in the Toronto Blue Jay's stadium. Could that be the answer?

Since it was the middle of the night in the UK, and this was twenty years before the Internet, I phoned the friend, Mo Dixon, who we had stayed with in Toronto and asked her to find out the address of the Sweetheart Company. Mo phoned back from Toronto University, where she was a professor, with the number in Pittsburgh. A phone call to them informed me that they had just started a company in England and that it was run by a man called Lionel Klacken. I found this astonishing, as I had done the catering for Lionel's daughter's wedding some ten years before, and we had got on extremely well.

I was even more astonished when I spoke to him the next morning and after a friendly chat over old times and the fun we had had over the wedding, and how I had left him a happier but much poorer man, he told me that they had imported the first Sweetheart machine into the country the previous week and it was being installed at that moment ready to start a trial at the NEC for the Motor Show. It was about fifty yards away from our pavilions where I was at that moment.

When people tell me that I have been lucky through my life, I always tell them the tale of the devout man trying to escape a flood by sitting on the roof of his house. He tells the boatman who rows out to rescue him, "Don't worry, the Lord will save me", which he also repeats to the helicopter pilot who is on the same mission. When he eventually drowns, and arrives in heaven he is furious with God and says "You said you would save me, why didn't you?" and the Lord said "I sent a boat and a helicopter, what else could I do?"

You need to recognise your luck for what it is, and allow it to work for you.

Three amazing coincidences allowed me to solve a problem that I didn't know I had had twenty-four hours before.

I suppose at this stage I ought to explain what exactly the Sweetheart Machine was. It was a 20-foot long conveyor, which took trays containing 24 wax paper containers, each larger than one pint (to allow for froth), and passed them under a series of pipes that filled each cup with an exact measure of one pint. At the end of the line a sheet of Clingfilm was drawn over the tray of full cups and then a hinged lid of heated rings was pressed down on the top of the tray, sealing the Clingfilm to the cups. This meant that we then had a couple of dozen sealed pints of beer, which remained good for 24 hours.

There followed a stroll across to the cellar where the Sweetheart Machine was set up to watch it working; a detailed discussion with the mechanic; a call to the Cornish weights and measures man and a visit by him to the NEC; a call to the current brewers supplying Twickenham telling them that if they wanted to keep the contract they had better buy one of these machines and, by the time, three days later, that I first met Bob Weighill, the RFU Secretary, with John Coomb, I had a fully-thought-through answer to their problem, complete with some sample beers poured the previous night at the NEC into Sweetheart Cups, kept fresh by their Clingfilm wrapping across the top.

It was quite a coup de théâtre as I first allowed them to take me round the ground and dwell at some considerable length on the enormous problem, which could not possibly be corrected until the stadium was rebuilt and more electricity allowed the beer pumps to be fully automated.

When we settled down in the Secretary's Office, I told them that I had the answer, showed them pictures of the machine in operation, offered them a copy of the letter from my new best friend from Cornwall confirming the legality of the new system, and finally presented each of them with a perfectly good pint of day-old Brummie Beer. The RFU Secretary, Bob Weighill, his administrative secretary, Colonel Dennis Morgan, and even John Coomb, thought I was a magician. I allowed them to think it – it would do no harm for our future relationship, which would turn out to be one of the most satisfying and profitable of my career.

Before I could get to this happy state of affairs, we had a very busy and extremely fraught three months work, preparing ourselves for this very different operation. It was now the end of October and we had to be at action stations for the first international, against Scotland, the first week of February.

While we had decided, along with the Cornish weights and measures man, that the Sweetheart Cup system was going to solve the supply problem for the bars, we still had to work out how exactly it was going to operate. Other than the brief trials at the NEC during the Motor Show, no one in the country had operated the infernal machine – and it was complicated. We were going to try out a totally new operating system supplying eleven bars around the ground, while also taking over an extremely complicated catering operation mid-season and open the new South Stand with 40 hospitality boxes, the first time the RFU had had them at Twickenham. I might have convinced the Secretary and Administrative Secretary of the RFU together with John Coomb that I was a magician; I now had to convince myself.

We worked out the Sweetheart Machine could produce probably a couple of thousand an hour and with the logistics of distributing trays of beer to the bars all around the ground, we decided that we would start the machine rolling at four in the morning and would have several thousand pints poured before anyone arrived at the ground and with this, together with the existing beer pumps in the bars, we would crack the problem.

What we didn't know was how many pints of beer we could actually sell, since the limitation of the previous system restricted sales to 12,000 pints per match – considering the fact that there were upwards of 70,000 beer drinkers at each match that seemed to suggest that we had some way to go.

The first match went very well with 20,000 Sweetheart Cups sold and an additional 4,000 pints through the normal pumps. This rose to 30,000 in total at the French match a month later. At the end of the first season we decided to add another machine on the other side of the ground and so for the Irish match we achieved the amazing figure of 60,000 pints – five times what used to be sold - but still less than one pint per visitor.

We were very popular with the punters, and the RFU, and were feeling rather pleased with ourselves – I should, like Caesar, have had someone behind me, whispering in my ear, "Beware hubris".

The Welsh match was next – the fourth home international and the completion of our first circuit of the Five Nations Championship. Several people had told me that if they lost, the Welsh always went home after the match. Having experienced the visits of the other three nations where, win or lose, everyone stayed behind after the match for a serious drinking and singing session, I considered that this was ridiculous. In any case, sadly, the Welsh didn't often lose to England at that time.

Sod's law being what it is, of course, they did lose this time and sure enough, off they all buggered back to the valleys to drown their sorrows in Brains or Felinfoel bitter, and even the exultant English supporters couldn't make up for their absence. I was left with 20,000 pints of beer that was going to be useless by the next morning, let alone the next match.

One of my more costly mistakes but mightily popular with all the rugby clubs in the area when our vans did a tour of West London delivering unexpected late Christmas presents of very palatable free beer.

12

Patients of Job

What I missed out when listing the hurdles we had to leap in the couple of months between Ring & Brymer quitting Twickenham and Paynes taking over was the not insignificant matter of my company coming within an ace of going broke!

I had, of course, known that the company was chronically short of capital, as my father, Gus, had looked upon the business as a means of enjoying his life, not of building a heritage. As a consequence, my inheritance from him a couple of years earlier had been somewhat of a mixed bag.

Gus Payne had bequeathed to me a very fine, very large, list of friends. The only trouble was that a goodly number of them were of the 'fair-weather' variety who had, over many years, revelled in my father's bountiful hospitality and given very little, if anything, in return. Partly as a result of this 'bountiful hospitality' and partly because Gus, as I have explained, liked to deal in cash, the company he left me had little except debts.

However, in business you are as rich as your creditors think you are, as long as you can pay their bills eventually, and Gus had given every appearance of considerable wealth. He was always immaculate, usually in three-piece suits with waistcoats, each with a different pattern of buttons in keeping with the various Royal Regiments of the Brigade of Guards.

He was five-foot-eight but always looked shorter because of his portly build. All his shirts were from Jermyn Street and every one had to be altered because of his short arms. He had, when he died, 60 pairs of hand-made shoes from Wildsmiths in St James's. He changed his footwear at least three times a day, which sounds indulgent but he had reason. His right foot was a good inch shorter than the other, almost a clubfoot. He said that changing them regularly was a necessity and it obviously worked because very few outside of the family knew of this peculiarity.

Since his shoes were always beautifully polished, kept in perfect order and never used for more than two or three hours at a time, we had potentially many thousand of pounds worth of beautiful footwear to get rid of after he died but my mother dealt with this and I discovered that she gave them to a charity. As they were hand made, each shoe looked the same size, whoever bought them must have thought they had had a fantastic windfall, until inexplicably their right foot lost all feeling.

One of my inheritances from him was an amazing collection of hats, all from James Lock, the Hatter in St James Street; traditional black bowlers and silk toppers, grey and brown bowlers and flat top bowlers, known as 'Cokers', in brown, black and grey – he really was quite a dandy. Immediately after his death, my mother suggested that I wear these hats but I pointed out that I had spent the last twenty years specifically not aping my father's style, as I did not want to be Gus Mark II, so I certainly did not intend to wear his hats. I said that I would keep them and wear them in another 20 years when they could be thought of as my hats – 30 years on I still haven't worn them, except for the silk topper for Royal Ascot. Given his grand appearance, it was neither surprising that everyone thought he was very wealthy, nor that in reality he wasn't. On Gus's departure, however I, at the age of thirty-three, had to keep up the appearance of being able to pay my way, although not nearly so extravagantly.

Progressively I weeded out the more obvious wastrels from the visitors to our events. However it was to be nearly three years after his death before we had the final hurrah of 'Gus's Freeloaders' at the 1981 Open Championship at Royal St George's Golf Club at Sandwich in Kent.

That was to be one hell of a week. It was the first time that the Open had come south for thirty years and very popular this was, particularly with our corporate customers. We were catering for some five thousand guests a day in sixty pavilions. In the Paynes entertainment pavilion I expected forty invited guests a day that I thought might go up to sixty.

The first day of the championships and, while everything was going well throughout the operation, the Paynes' tent was being deluged with the 'old brigade'. By one o'clock we had served over a hundred for lunch and more were arriving by the minute, all of them feeling they had the congé becoming to a good old pal of the late lamented Gus.

Judy was helping me deal with this mêlée and, while dashing around like a whirling dervish, I noticed that one of our great friends on the golf circuit, the Australian Graham

Marsh, brother of the Aussie wicket-keeper and known to the golf world as Swampy, had arrived with a friend and was being given a drink. Judy joined them as did I several minutes later when I could extricate myself.

"Hi there, Payney!" said Swampy. "Let me introduce my caddie for the week, Vice-Chancellor of the University of Western Australia in Perth, Don Watts",

"Delighted to meet you Don, you're very welcome",

"Well, Alan, it sure is nice of you to take the time out to greet us and have a drink when you're so obviously busy – although I suppose by now everything is so totally out of control, there's fuck all you can do about it anyway." The laugh that accompanied this was somewhat like a kookaburra.

Quite an opening! I replied through a gritted smile, "Don, please keep it quiet, they still think I'm in charge."

All this, of course, was some time in the future, but the problem after Gus died in November '78 of keeping up appearances, while reducing costs, was very real. I obviously must have succeeded as the company expanded quite dramatically over the next couple of years, not least with the arrival of John Coomb and the addition of Mayfair. I thought that we would ride out the lack of capital because, in my naivety, I didn't realise that many more companies go to the wall through too much business than too little.

We were over-trading like mad. A three-fold increase in business, new premises, which had to be completely refurbished, all in the space of a couple of years, and we had had no increase in capital. What's more, we were about to take on a contract at Twickenham that, while immediately very limited had, I believed (with justice, as it turned out) the potential for enormous growth.

This all came to a head, ironically, on the December

day we visited Twickenham to see Ring & Brymer operate on the last day of their contract, the Varsity Match 1980. Though I had known during our busy and successful summer we were progressively having difficulty in paying our suppliers and were perilously close to our limit at the bank, I did not realise how close we were to being insolvent.

I had a call from Leeds on the Tuesday morning of the Varsity Match from John Murtland, my accountant, who said that he had some serious business to discuss and would be arriving at Kings Cross at eight o'clock that evening. He would probably need all my time for the rest of the week. He briefly explained that we were running, pell-mell, into a serious cash-flow crisis and needed to move very quickly to avoid a catastrophe.

As you can imagine it was a very chastened, not to mention worried, Alan Payne who went to Twickenham later that morning, although it was absolutely essential that no vestige of the problem should appear on my face. No matter what the position, it would not be helped by the RFU worrying about the viability of their new about-to-be-appointed caterers.

Either through good luck or, more likely, hard work, I was able to keep up a happy demeanour almost regardless of any inner turmoil – in other words I had developed a "jolly shell". Since my whole working life has been predicated on helping other people to enjoy themselves, and this had grown into a fervent belief that customers do not want to see a po-faced caterer. I have done my best over forty years to ensure that they don't. Whenever our customers, at functions or events, commented on how well or cheerful I looked I usually responded, "It's just a façade - it crumples when I get home" – it was only partly in jest.

This ability was seriously tested on that Tuesday, for

as well as the RFU, I also had to spend the day with Ring & Brymer as hosts, showing me around the facilities along with my head team. My rivals were very far from happy about the decision from on high that they should withdraw from Twickenham and would, of course, have delighted in passing on to the world at large the wonderful news of Paynes Caterers' imminent demise. At least I could dispense with the false bonhomie and pretence that had been my necessary companions all day when I met John Murtland from his train that evening.

I had known John and his family all my life. Gus had been stationed with the Royal Engineers near Leeds during the war and he and my mother lived in a village called Aberford. It was in the local pub, the White Swan, that he met John's father, Bernard, also an accountant. They played dominoes together and when Gus needed someone to do the books for the Sergeants' Mess, Bernard was the obvious candidate.

Incidentally, I was conceived during this period, although my mother had the presence of mind to head back south in the latter stages of pregnancy, and I was born in my aunt's flat in Oxford, three weeks early, with my first night in a drawer in my sister's doll's clothes. I was, however, safe in the knowledge that I could, of course, never play cricket for Yorkshire – for which Yorkshire has been eternally grateful.

When, after the war, Gus set up his business in Shepherds Bush, Bernard became his accountant and auditor. This was based on a wonderful arrangement. Bernard was very fond of watching tennis and, through his friendship with Fred Perry (who used to practice at the Lyons Club at Sudbury) Gus had access to tickets for Wimbledon. Since both parties subscribed to the ancient system of bartering, the audit naturally took place each

June during the championships and included a couple of visits with an elegant tea on the members' lawn. This very civilised arrangement continued for the rest of their lives. John then took over from Bernard as I did from Gus. Sixty years on, but without the Wimbledon tickets, John still looks after my affairs, the financial ones anyway.

I said that people do not expect to see a po-faced caterer, but that is precisely what they do expect in an accountant and John certainly did not disappoint. He has a wonderful 'dour-Yorkshire-accountant's face' when he wants - I am pleased to say he has an awfully jolly 'Tetley-beer-drinking-face' as well.

In fact that was the only thing we seemed to disagree about, beer. Since he was from Leeds he was brought up on Tetley's with the occasional glass of Sam Smith. He was very disparaging about our local London beers on one occasion quite early in our friendship when I had taken over the accounts of Paynes Caterers after the sudden death of our finance director.

John had just joined his father's firm in Leeds and we were his first away visit. Because Bernard had always stayed with Gus and Lila while he did the audit, it made sense for him to stay with Judy and me, so starting a second-generation tradition.

At that time some awful beers, such as Watney's Red Barrel, were being foisted upon the beer-drinking public and unfortunately this particularly dreadful one was made at Mortlake Brewery, a couple of hundred yards from where we lived at that time. Luckily there was also a local brew of which I was very proud, called sensibly enough, London Pride, and so John and I found ourselves in a pub owned by Fullers, the producers of this brew.

After taking one draft of Pride, John said in his flat

Yorkshire brogue: "It's a bit luv-in-a-punt is this." I said, "I'm going to regret asking you this, but why is it 'a bit luv-in-a-punt?'" "Because it's fucking close to water." I have yet to hear a better derogatory description of a beer since – but totally wrong of course.

However, it was very much the 'dour-Yorkshire-accountant's face' that I saw heading off the platform at Kings Cross. He had good reason.

The tale he brought was very gloomy – we were insolvent and unless we got a healthy increase in our borrowing capacity from the man at Barclays, together with a breathing space, we were going to run out of cash. Quite simply we didn't have the capital we needed to grow as we had been doing and we would probably have to call in the receivers. Pretty serious stuff.

It required me making an appointment for the end of the week with 'the man from Barclays' who was, in fact, the local director, together with the manager of our branch in Hammersmith. Before this meeting at the local head office in Chelsea, John and I spent three days putting together a very detailed business plan, justifying why they should allow us more credit. As with all business plans it required a great deal of 'crystal-ball-gazing' and would succeed only if the man receiving it believed what he was reading.

This was where my assiduous use of our events for entertaining those who matter proved so critical. I had invited the Barclays' local director to see us in action on several occasions, and since this involved the not disagreeable task of eating great food and drinking exceptional wines in the company of some very interesting people, usually at an event where tickets were at a premium, it worked quite well. Obviously the lunch at Muirfield earlier that year with Matt Busby, and the dinner at Glenconnor, highlighted to him that the assets of the company did not all appear on the balance sheet. He had been able to see the big picture and, come the day of the pow-wow, it all paid off as the local director gave me the go-ahead but with quarterly reviews – I was on trial, but I could proceed with the Twickenham contract.

This was brilliant and saved the day but my problem subsequently was the local manager who resolutely refused to see the big picture. This was not really surprising as I was far and away his biggest 'problem customer'. Since his personal sanction was certainly less than ten thousand pounds and I was borrowing something like three hundred thousand, and wanting considerably more, you could see his dilemma.

Sadly he had not been at the lunch at Muirfield with his boss and so he firmly believed that the company was a one-man-band - so much so that I felt he wanted me to speak to him every day to prove I was still alive. He also insisted on additional insurance on my life, which, together with all the policies we had to take out when we bought British Grove meant that the best thing for Judy ("Payne-the-Purse") would have been for the warehouse to burn down, preferably with me in it. I made sure she wasn't let loose with any matches during that time.

The almost micro-managing by my local manager became more than a little irksome during the very busy spring and early summer of 1981. With the first two outstandingly successful internationals at Twickenham, the preparations for a very strenuous summer season underway and John Coomb continuing to bring home the City bacon, the last thing I needed was a lack of faith from my bank manager.

However, among that mêlée in my pavilion at the

Open at Sandwich in July 1981, Martin Devetta, who had put up all the tents, brought his bank manager in and I gave him lunch. This banker, from Natwest, seemed a very different kettle of fish to my guy and, since he was from a City of London branch, he was not fazed in the slightest by the borrowings I talked to him about, which was a lot less than some of his commodity dealers dealt with as small change on a daily basis. A couple more meetings and some detailed information and we agreed that we would move our account to his bank.

I then had the very pleasant task of telling Barclays, who had become progressively more tiresome during the summer. I was summoned to Chelsea for my quarterly review and as we sat down I said that I hadn't brought any figures with me as I was moving the account.

The effect was just as wonderful as I expected and while enjoying it I was reminded of the story told to me by our very good friend, Mo Dixon. Her uncle had owned a very large greetings card company, which quite naturally, developed a considerable overdraft in the run-up to Christmas. One day he was peremptorily summoned by his bank manager in the Yorkshire town of Keighley and told, "Dixon, I'm very worried about your overdraft." The brilliant response from Mo's uncle was, "Nay Lad, there's no point both of us worrying on it, I'll tek it elsewhere". And he did, and I did.

My new 'Man from Natwest' rejoiced in the name of Brian Pay. Imagine! What a great name for a banker! He was a great chap too and while the perennial problem of under-capitalisation was going to take a further eighteen months to really sort out, he was a very positive friend to have on my side, continually complaining in a droll way about the problems of "growing Paynes".

One visit he made in that first winter was as my guest to the England v Australia match at Twickenham and it helped considerably. As he sat in the front row with a client of his, he was somewhat flummoxed when, at half-time, five feet away, a very handsome young lady called Erica Roe stripped off her clothes and jiggled her considerable attributes right in front of the two of them. Brian said afterwards that he was enormously impressed both with my organisational ability and the novel lengths to which I would go to secure my overdraft. It certainly worked.

With stalwart support from Brian Pay and some very effective management consultancy from John Murtland's London partners at Deloittes, we worked over the winter of 1981 and through the spring of 1982, repairing the ravages that lack of investment over twenty years, together with very rapid growth, had inflicted upon the company. We also put in place a very sophisticated reporting structure capable of handling the very difficult operational needs of a growing event and function caterer.

However, by the summer of 1982 it became obvious that we were not going to be able to continue without some serious long-term financial support. The irony was that I was doing business on a regular basis with any number of people who could have supported us, but I couldn't let it be known that I needed it. I had to wait for them to come to me.

On the evening before the Dunlop Masters Golf Tournament at Woburn Golf and Country Club in June, Judy and I were included in a small party of Dunlop people invited to dinner at Woburn Abbey by the Marquis and Marchioness Tavistock, which included a personal guided tour of the magnificent building by our hosts. Quite an evening, made even more interesting by a conversation with

Graham Marsh, one of our fellow guests, who said that he had some money in a bank in the UK and would possibly be interested in investing in Paynes Caterers.

After the event, he continued his peripatetic life while keeping up a correspondence with me covering all aspects of the company's plans and requirements. A month later we had dinner at the Open at Royal Troon and he said that having thought the thing through, he really believed that, since he came to the UK only a couple of times a year, it would be too much of an arm's length involvement. I was not surprised and had considered his approach had a less then even chance of coming off.

Because of this I continued quietly to explore other avenues, including a meeting with Sir Stanley Grinstead, the chairman of Grand Metropolitan plc and one of the more reliable of Gus's old friends. Grand Met had been setting up an outdoor catering wing and certainly were very interested in acquiring Paynes, but before I had a chance to pursue it further a totally unexpected approach suddenly appeared.

I received a phone call from an old friend and client of John Coomb called Bruce Dehn. A lawyer straight out of Dickens, stooped back, parchment skin taut against his high cheekbones, never without a cigarette, he was, among other things, Clerk to the Worshipful Company of Distillers, which I was applying to join.

After some amiable banter and discussion about a recent dinner we had done for him he said, "Alan, are you interested in having someone invest in your company?" "I could be, tell me more," I said, trying to appear only moderately interested – I never was a poker player. "Well, a very good friend, who coincidentally is also applying to join the Distillers, by the name of Simon Roberts, Chairman of Job's Dairies, is very keen, can he contact you?" "Of course. Since I am an outdoor caterer, I'm usually outdoors, particularly during the summer so it's probably best to get him to give my secretary a ring and we can arrange for him to come over to British Grove for lunch in the Pantry." I always liked seeing people on my terms and on my patch.

So the next day Jenn, my secretary, received a call from Simon Roberts asking when the best time was to ring to speak to me.

"Best thing is early in the morning, Alan is usually here in the office before the day starts in earnest." "That's great," said Simon "I'm an early bird. What time does he get in?" "Oh, normally between five and five thirty." "Great Scott! I'm a milkman and I'm not up at that time".

So Simon phoned me at seven thirty the next morning on my direct line – because he didn't believe Jenn – and we arranged for Simon and his finance director, Bob Culver, to come to British Grove in a week's time, at the beginning of August, by which time we would have the results of the Open Championship and a detailed forecast for the balance of the year.

I had met Simon once before, very briefly a year or so previously, when Alan Cox, as our private catering director, had looked after a party to celebrate the golden wedding of Simon's parents. Following that party, we had changed our milk supply to Job's Dairy and, unbeknown to me, had been identified by Simon as a future investment. He had then waited until the perfect moment to make his move.

I was waiting in the yard at British Grove, wondering what I was going to make of these potential partners and was duly impressed when Simon drove in at the wheel of his beautiful old Aston Martin, with the roof down and Bob Culver by his side. The inevitable small-talk about the car preceded our entry into the Pantry and the wary initial sussing-out of each other over the first of very many glasses of Clicquot we were destined to have together over the next decade and a half.

They were both about the same height as me and both, at that time, a good stone or two heavier than I was. Simon certainly carried his weight considerably better than Bob and it was no surprise when Simon told me that he had only given up rugby a couple of years previously at the age of forty two. He had played quite seriously at school and for his club and only just missed getting a blue at Oxford. He naturally questioned me a great deal about our contract at Twickenham.

The arrangement was obviously 'soft cop, hard cop' with Simon softening me up with his considerable charm and Bob asking the hard questions, not surprisingly mainly on the figures. This did not come as any great shock as after twenty years in business I was not unversed in the ruses involved in negotiation. John Coomb and I often played similar games with clients.

We had a very enjoyable lunch at which I discovered that the Roberts family owned Job's Dairy and Simon was the fifth generation to run the company. Job's had expanded away from just owning the very profitable and cash generating dairy in Southwest London and now had subsidiaries including Southern Fast Foods, a franchise of more than fifty Kentucky Fried Chicken shops and Job's Fast Food Distribution, a fast food supply company.

I found his company at the lunch in our pantry engaging and was impressed when I asked him: "Why are you limited to South West London? Why, for example, don't you have any dairies in North London?" Simon's response told me a great deal about how he ran the company, "I don't like driving round North London."

Truly admirable – a perfect answer. I used a variant many times in later years when my fellow directors kept on at me that we ought to take on contracts at various questionable football stadiums. There is no point having a private company and doing things you don't enjoy - not if you can make money while avoiding them.

Another fact that impressed me was that their company newsletter was called "The Book of Job" which appealed to my love of word play and I realised by the end of lunch that, with a transfusion of much-needed capital we were shortly to be "Patients of Job". So started two new relationships, one with Simon Roberts, my backer, and one with Bob Culver, his henchman, both of which were going to be an important part of the next decade and a half, one totally positive, the other less so.

It was agreed that in return for the capital necessary to sustain the company and fund the considerable growth we anticipated, we would become a majority-owned subsidiary of HA Job Ltd leaving me with a substantial minority stake in a very much more valuable company.

I was to continue to run the company as managing director; Simon was to become chairman in place of Bill Graham and Bob Culver to be deputy chairman and to take responsibility for Paynes Caterers (which changed its name to Payne & Gunter at the same time as the takeover) on the Job's board. In other words, Bob was to be the conduit of information between the two companies.

On the day of completion with Jobs, I had one bittersweet function to undertake because John Murtland was going to have to stand down as auditor under the new regime. I wasn't going to let forty years go un-remarked or un-rewarded so I presented John with an agreement duly signed by me and witnessed by Simon Roberts and his brother Christian, which gave John access to any of our events:

This agreement

Between John Campbell Murtland, FCA, on the one hand, (hereinafter called the valued friend), and Alan Richard Payne, (hereinafter called the happy caterer).

Wherein it is here agreed that as a gesture of the esteem which the happy caterer holds for the said valued friend and by virtue of the assistance accorded by the family Murtland to the family Payne in war and in peace over four decades, the happy caterer does extend to the valued friend an invitation in perpetuam to call on the happy caterer at any golf tournament, tennis tournament, rugby match or other sundry event at which he may be catering so that suitable refreshment including stimulating liquors and the like may be provided as the valued friend may desire.

Given under our hands severally and jointly on this tenth day of January one thousand nine hundred and eighty three.

Alan Richard Payne *John Campbell Murtland*
Witness: Simon Roberts *Witness: Christian Roberts*

During the subsequent dozen or so years that we continued to own the company, John only took advantage of it a couple of times a year, notably at the home internationals at Twickenham and when the Open Championship was in the North of England. It was a very trusting and generous thing for Simon to do on taking control of the company. But then, Simon Roberts is a very impressive man, very shrewd and patient. I have always greatly enjoyed his company and have used him as a sounding board both during the time we were associated through our joint ownership of Payne & Gunter and even more subsequently. He is also a past master of the art of chairmanship, stemming, as he says, from his innate laziness. Simon is, in terms of my respect, up in the same league as Gus, Bill Graham and John Coomb.

At around the time of the Rose Room dinner we had a real disaster, which proved the wisdom of having a senior partner like Jobs Dairy, a company called Febbs Management Ltd, which ran a golf tournament called The Bob Hope British Classic, went into receivership among various tales of accounting irregularities. Peter Dobereiner, the wonderfully witty and literate golf correspondent of the Observer had reported these irregularities prior to the collapse but no one wanted to listen. After the event the case was taken up by Private Eye, resulting in a High Court libel action at which, to my chagrin, I was an exhibit.

The total the creditors were owed was more than half a million pounds, included amongst them was P&G who were owed forty six thousand pounds, of which we never saw a single penny. One of the main casualties was Martin Devetta's company, Spandrel Hire, who were owed ninety-two thousand. Because it was at the end of their season, their bank, Natwest, whose manager Brian Pay had moved on a few months earlier, took the opportunity to pull the plug and put Arthur Andersen in as receivers. Since the receivers ended up paying pound for pound to Spandrel's creditors, the receivership was a totally unnecessary panic move. The Bob Hope Classic and a shortsighted bank manager had destroyed a very fine business. I knew that had we not become Patients of Job earlier that year the same fate, with the same bank, would certainly have befallen us.

Since our contract for tentage at the Twickenham matches during the forthcoming winter was the only one Spandrel had on their books at the time of their receivership,

David Buchler, the Andersen partner responsible, tried to put the squeeze on me, stating that the price was going to be double that previously agreed, for the remainder of the season, or the tents would be taken down before the International which was to be played the following weekend.

I pointed out to David that perhaps he would like to phone the Chairman of Natwest (who after all were his employers in this, and many other, receiverships) and explain to him why he would not be able to have lunch with ex-Home Secretary Merlyn Rees MP before the match as he had planned. Needless to say the prices remained static.

Just after the debacle of the Febbs' receivership, when it became obvious to all the creditors that nothing was going to be forthcoming from the ashes, I wrote to Bob Hope personally, and, more in sorrow than anger, said that it was a shame that his name, which had, quite reasonably, been taken by suppliers as a guarantee of the probity of the event, would obviously be tarnished by other well-known companies collapsing as a result of this failure.

It was, though I say so myself, a bloody good letter, but it drew not one whit of response. I wasn't really surprised, as the whole story of the tournament, which was supposedly for charity, had been surrounded by tales of egregious expenses by Hope and his cronies and very little for the charities.

Later that year at the Open Championship at St Andrews my son, Richard, and I were having dinner with Graham Marsh and Hale Irwin, the US Ryder Cup player and former US Open winner, and told them about the Bob Hope Classic and my letter.

Hale said, "I know Bob Hope very well and he really is a good guy and obviously the letter never got to him personally." I said, "Hale if you know him that well, can

I give you a copy of the letter for you to hand over to him personally so that you know that he has read it and we can both know if he's a good guy or not?" "Sure you can." And I did and he did and we do.

It was after this, speaking at a golfing dinner, I made light of it by suggesting that over the gates of Moor Park Golf Club, where the tournament had taken place, they should erect a sign: "Hope abandoned all who entered here".

13

HORSES FOR COURSES

While for Gus horse races were an enjoyable pursuit, which ate up a certain amount of his free time and free cash, I did get a bit more involved in the 'Sport of Kings' as a business proposition, although my first serious outing was far more of a gamble than Gus ever took with William Hill. Not long after the start of the partnership with Simon Roberts and Jobs Dairy, an old drinking pal of Gus's, a wonderful buccaneer called John Hughes, approached me. John was the Clerk of the Course and director of three or maybe four racecourses, the representative in the UK for Waterford Crystal, and a director of an Arab-owned Scottish mineral water company. These were the roles that I had known about but, since he prided himself on having thirteen hats, there must have been some others.

One of these missing ones he mentioned when he came to see me in the late summer of 1983. It seemed he was also a director of Racecourse Holdings Trust, known to the racing world as RHT. This company, owned by the

Jockey Club, the aristocratic organisation that runs racing, held shares in various racecourses and in a tax-efficient way used the profitable ones to support the less successful ones.

The deserving cause at that moment was Aintree, home of the greatest steeplechase in the world, the Grand National. RHT had just purchased the course from the Topham family and had signed a new long-term sponsorship deal with Seagram Distillers and so had assured the continued existence of the race. John, as Clerk of the Course at Aintree, (the man responsible for the racing), had been very much the driving force of the movement within racing to save the National, and it was, of course, about Aintree that he came to see me.

John wanted to involve us in the National as a way of bringing a new big player into the world of racecourse catering, which at that time was very largely shared by Ring & Brymer and Letheby & Christopher. He also had thoughts of tapping into our proven ability to market packaged hospitality, in other words to bring some of our golf, rugby, rowing and City of London clients across to racing. After a preliminary drink in our pantry at British Grove - the first of many - when he told me of his ambitions for Aintree and the opportunities that waited for us in racing, I arranged to go up to Liverpool and meet him at the course.

When I, together with a couple of colleagues, arrived at Aintree, John firstly introduced us to the general manager of the racecourse, Rod Fabricius, and little did I know how important, and what a good friend, he was going to be. John had told me beforehand about Rod, who had been very much his protégé in various roles in racing prior to this one.

They obviously had enormous respect for one another and worked well together but they made a contrasting pair.

John, usually in tweed suiting with thick horn-rimmed glasses on his nose, somewhat jowly, with a gruff voice and ready wit, was a 'hail-fellow-well-met' sort of character. He had a very ruddy complexion that was possibly caused by too much bitter wind on racecourses or, more likely, by too much windy bitter.

Rod, on the other hand, was tall, very slim, usually in either well-tailored slacks and tidy shirt or, as when I first saw him at Aintree, in a fine lightweight dark-grey suit, blue shirt and tie. He was always extremely elegant with somewhat saturnine, handsome features, flecks of grey in his curly black hair, and a somewhat preoccupied, serious look about him. He also had a very dry sense of humour but he hid it well, very well.

The only thing that Rod and John Hughes had in common in appearance, was that they both wore the badge of office of the horseracing man, a brown trilby, battered, with the prow well worn by the greasy thumbing caused by the hat being formally raised every time it was in the presence of ladies, owners and, of course, members of the Jockey Club.

Rod took himself and his job very seriously and as he showed us round it became apparent just how much he and John and the RHT had taken on. The facilities were abysmal and had not changed since the buildings were put up in the eighteen-nineties. We were going to be hard pressed to make our mark by improving the catering, given the conditions we would have to work in.

When we returned home and started to put together our proposals for the catering operation, I decided that John was right, the most advantage would be gained by making use of our well developed, and increasing, ability to market packaged entertainment to a growing database of satisfied customers. What we had successfully done at Henley and Twickenham we hoped we could do at Aintree. However, our customers were used to facilities far better than those available for the National. Since the right facilities were not there, we were going to have to create them.

I had been having some discussions with Ron Millward, the grandstand builder who had initially introduced us to Henley, and who was attempting to get someone to try out a new type of facility he had designed. This was a grandstand made of scaffolding which incorporated a line of glass-fronted pavilions on the ground floor with a section of raised seating above and a glass-fronted restaurant for two hundred on the top.

Since, during the subsequent twenty years this structure, and many variations of it, have become routinely gathered around virtually every sporting arena or track as a way of generating income for sports, it is difficult to realise just how radical and risky it was to put the first one up, sell it to a sceptical clientele and operate it, learning all the while.

Most importantly, if we were to go ahead, we would be agreeing to pay for the erection of the structure with no guarantee that it could be sold. Without our new shareholders from Jobs there would have been no way that we could have taken it on. Even so, it took a great deal of persuading on my part and some considerable trust on theirs, but we established a major new facility for the corporate hospitality industry.

Coupled with the fact that this was our first foray into racecourse catering, with all the problems that entailed, we had to contend with the inevitable delays in the building of the new structure. We also knew that we were going to have to set up something totally new inside it – but only once it had been completed. Since Ron Millward and his team doing the building were learning whilst the edifice was growing, it was a nerve-racking few days.

The site for this new structure that we were given at Aintree was round the corner from the main grandstands, immediately alongside the start, where upwards of fifty horses and riders jockey for position before the tapes are raised for the beginning of the race. A wonderful location and, although we agreed that for Grand National Day itself, the stand would carry a banner for the sponsors, Seagram, on the first day it would carry one for Payne & Gunter, in full view of the television cameras.

When the company name was changed to Payne & Gunter, which coincided with the Jobs' deal, we undertook the only advertising campaign in the company's history. It comprised full-page adverts in the Sunday Times Colour Supplement and was based upon a series of images of different sports and events, each one with the strap-line, "Everything Goes Down Rather Well With Payne & Gunter".

It was therefore this slogan that was brandished across the top of the new structure on the very first day that it had ever been used. The night before, I woke sweating, from a vivid dream that at the start of the first race, everyone moved to the front of the structure to see the start and as the stand collapsed the last thing the television cameras captured on the descending chaos was the sign fluttering across the wreckage, "Everything Goes Down Rather Well With Payne & Gunter".

Fortunately, for Aintree and myself, my dream was not realised, the stand did not fall down. We had managed to sell out both the pavilions and what we referred to as the 'Restaurant in the Sky' for the main Grand National day. However, we only partially sold the other two days and so only just broke even on the investment. Still, not bad for our first attempt and a great deal of good publicity.

Sadly the same could not be said of the rest of the catering where we were not able to make the improvements we would have liked, not only because of the dreadful limitations of the facilities but also because the Aintree directors would not let us make the wholesale changes in the staffing that we wanted. The trouble was that, while there had been frequent changes in catering contractors over the past few years, each new caterer had continued with the same staff, and therefore the same dubious practices, which had become enshrined in stone.

Liverpool people, scousers, (named after their Irish or scouse stew), have a delight in being at odds with the rest of the country. The reputation of being both very funny with a quick, sardonic wit, but also feckless and work-shy, has developed over many decades and at the time we arrived at Aintree in the mid-eighties, the local council under the leadership of the sharp-suited socialist, Derek Hatton, was doing its best to reinforce the latter view to the world outside.

Prior to doing the Grand National, my only real exposure to Liverpool people was through Jimmy Kirkham, the chef who had started Paynes Caterers with Gus and who had stowed away on a liner from Liverpool docks at the age of thirteen. Spending many evenings during my first years with the company listening to Jimmy's wry tales of his Merseyside childhood gave me a background belief in good-natured native scouse humour that sent me unwary into Aintree. While Royal Ascot in the early sixties had been quite something, and the subsequent twenty years' exposure to the seamier side of the catering trade had certainly hardened my skin, nothing prepared me for what we found at the National.

The whole week was considerably more difficult than anything we had experienced before, culminating in the nightmare final afternoon as, totally unannounced, the security guards and police all appeared to go off duty as soon as the Grand National had been run. Since this obviously had been the time-honoured practice at Aintree, all the local lads were ready and waiting outside the course until the race was over. As the winning horse and jockey were, as always, led by the Chief of Police of Liverpool, on horseback himself, dressed in his formal dress uniform with a plumed hat, into the winner's enclosure, the Liverpool lads descended in swarms, like locusts.

Dozens of them made their way into the stands and particularly the marquees and since many of them were certainly in cahoots with some of our casual local staff it became impossible for our own staff to control the situation as cutlery, china and booze was taken and carried away to cars parked in readiness. It was a well-planned raid and in short, we lost ten thousand pounds worth of liquor, twenty-seven thousand pounds worth of equipment and eight hundred pounds in cash from a till. It was a very costly lesson.

A pleasing coda to our first trip to Aintree abolished in my mind the conviction that all Liverpudlians were work shy; we had a group of young guys, who came to work for the week as porters, were absolutely great and worked very hard. After the race meeting they followed us back to London, and remained a vital and hilarious part of our team for the next twenty years.

We decided that we would have another go at Aintree the following year; putting into practice the lessons we had so expensively learned. But our hopes for the future at Aintree took a severe knock when we discovered that Rod

Fabricius, with whom I had got on so well, had been offered the job of Clerk of the Course and General Manager at Goodwood Racecourse and we would have a new untried boss at the National. It didn't bode well.

What we were going to do, besides getting the new General Manager at Aintree to agree that the security operation should last until everyone had left the ground, was to instigate our own staff frisking operation that had worked so well at other events. Quite simply, we set up a chicane of barriers that all staff had to come through on the way to the cabins where they received their pay at the end of the event. In these chicanes our security staff searched the staff and their belongings. What had happened on previous occasions at the Open and elsewhere when we used this technique was that the staff who had drink or equipment hidden about their persons, when the message about the searches came back down the line to them, disappeared into the nearest toilet and left their loot which was then collected by other Payne & Gunter security people stationed in the loos.

At other events this method had reduced our losses dramatically and we had high hopes of it working at Aintree. However it had one flaw, one serious flaw. It ignored the quick-witted nature of the scousers, who, when the message came back down the line that they were going to be searched and any booze confiscated, decided that the obvious thing for them to do was to drink it there and then. This they did, resulting in a drunken mob rioting and almost turning over the pay cabin with my wife Judy, among others, in it. Our team were all right afterwards due to the application of medicinal alcohol themselves. Needless to say the Aintree security and police had, as usual, disappeared after the National, ignoring the half-hearted instruction of the new General Manager.

After much soul-searching we decided we would go back to Aintree for a third year. We had many months' negotiations with the Aintree management and had finally settled the extraordinarily detailed terms of who would pay for what and how much and were about to sign a new contract. At this point Bob Russell went up to the course on a site visit and was suddenly told by the General Manager that Aintree had decided unilaterally that, regardless of what the contract said, we would now have to pay for our own staff toilets. The unnecessary and unexpected cost, but even more the breath-taking arrogance, told us very clearly that other costs would unquestionably be added at the whim of the Aintree management, and so made the decision to withdraw inevitable.

When I announced our decision to the P & G team I got the biggest cheer of my career, but at least one good thing had come out of our second Grand National. One of the visitors to have lunch in the Restaurant in the Sky was a man called John Prudence, the Managing Director of the Goodwood Group of Companies and as such, under the owner Lord March, the new boss of Rod Fabricius. He was very taken with the structure and our enterprise in setting it up and filling it with guests at a very high price. It was to reflect well on us a couple of years later when we bid for the contract at Goodwood Racecourse.

Goodwood had always been my favourite racecourse since Gus sent me to work there for Ring & Brymer in 1962, as part of my learning about the wicked ways of outdoor catering. I had driven down to Sussex in my grey mini-van accompanied by my old school friend Peter Challands, who wanted to earn some money before going off to Cambridge that autumn, and another character who would be an important figure thirty five years later when we re-emerged at Goodwood as the new caterers, Neville Petts.

Neville was several years older than me; he was an hotelier primarily but had joined Gus to work for the summer season as a junior manager. Outdoor caterers, then as now, provide a useful source of income and training for hotel and catering managers either between jobs or as holiday bonuses. He and I had got on from the start and so commenced a forty-year friendship that has survived quite a few storms, as neither of us could be considered easy-going. What is that conjugation? 'I am strong-willed; you are stubborn; he is bloody-minded.'

We arrived at Goodwood in the summer of 1962 and reported to John Coomb, at that time the director of Rings in charge of the Goodwood contract. I was put in charge of the sandwich room, where a wonderful and very funny group of ladies from Doncaster spent all week making sandwiches for all the bars and snack points. They took me under their voluminous wings and made sure that the bar staff didn't take me for too much of a ride. Peter became a porter in one of the bars and Neville was put in charge of the catering operation on Trundle Hill which was so steep that he swore he ended up with one leg shorter than the other. It was, as Royal Ascot had been, an object lesson in the fiddles of the trade and helped my education enormously.

Of all the many tricks of the bartender's trade that I learnt at Goodwood that week was one, which, for sheer chutzpah, has remained my favourite. The bar would as usual have a couple of gin bottles upside down on optic stands ready to dispense the correct amount once the glass is pushed up against the measure. However, only one contained gin. The other had, once emptied, been refilled with tap water. When a punter had had his first few gins from the real bottle he was primed and ready to be conned. Under the counter, out of sight, was a saucer with gin in it and the rim of the punter's glass was dipped into it before being served a measure of water from the doctored bottle and tonic sloshed onto the top. With the taste of gin on the rim and the bubbles of the tonic, I defy anyone to tell that he was being fleeced – until, after a long afternoon's drinking, he is still sober.

Overnight we slept in one of the bars in the Victorian grandstand and the feeling when we got up in the bright sunlit mornings and had the magnificent vista from Goodwood Hill all to ourselves made me totally fall in love with the place. Subsequently I went to the course many times as a punter, with Gus, but always had an ambition to take over the catering.

I had therefore, during the subsequent twenty years, looked longingly at Goodwood, knowing that it was, without question, the racecourse that most suited the market in which we saw ourselves. Described by King Edward VII as 'a Garden Party with racing tacked on', it was stylish, upmarket, and slightly aloof from the common fray, which is precisely what Payne & Gunter aspired to be. Goodwood remained, however, tantalisingly out of my reach as a potential caterer for the whole of that period.

By the summer of 1984 things had changed. Our first sojourn at Aintree, had, quite evidently, proved that we were capable of joining the ranks of caterers prepared to lose money on racecourses. I had established a good working relationship, not to say friendship, with the man who was Goodwood Racecourse's new General Manager, and we had shown ourselves to good effect to their Managing Director. It was not a bad position from which to mount a challenge to the existing caterers, who incidentally were helping us considerably by being less than caring about their standards at Goodwood.

It was at this time that another of those lovely coincidences that have peppered my life occurred. At a trade wine-tasting, a friend of many years, the enormously agreeable Chris Purchase, whose family had been selling wine from their delightful bow-fronted Georgian shop in Chichester for several generations, told me that he had heard that the caterers who ran the restaurant at Chichester Festival Theatre were going to retire at the end of the summer season and the theatre would be looking for a replacement for the following summer.

Although restaurants as such were not our normal activity, we had been joined that spring by my old friend, and champion of Trundle Hill, Neville Petts, who, having had a very successful twenty years working for various hotel and restaurant groups had been persuaded to come and add some professional large-company management to our rapidly growing business. So, with the appropriate skills available in Neville, the theatre restaurant seemed too good an opportunity to miss; hopefully it would be profitable on its own account but it would provide a wonderful way to get a step closer to Goodwood as Lord March had been involved in his local theatre since its inception.

We decided to rename the restaurant 'Gunters', largely because we thought that the name would strike a chord with the ageing clientele of the theatre, many of whom would have doubtless been taken to Gunter's Tea Shop in the West End by their aged aunts or grandmothers. It also would subtly infiltrate the company's name into the area. Neville and my wife, Judy, set to work to transform the rather decrepit cricket pavilion into a very stylish restaurant for the start of the 1985 summer season and we then ran it with modest success for the next four seasons.

During our first season, the theatre, as part of their ongoing relationship with the Goodwood Estate, were made title sponsors of a race at the May meeting at the racecourse. All the stars of the season's productions were invited to the race meeting and were entertained to a lavish lunch in a large room in one of the old Victorian buildings that were soon to be replaced. Judy and I were invited with the theatre guests and, being nosy caterers, during the afternoon we strayed into an adjacent room, where a very elaborate buffet was being set up and photographed. Not knowing who we were but assuming we must be theatrical people, the photographer asked us if we would mind standing by the buffet with a glass of wine to add some atmosphere to the pictures.

We agreed and obviously we enhanced the buffet so much that the photograph ended up on the front cover of a new brochure for our major rivals, Letheby & Christopher, the existing Goodwood caterers. Sadly, at the very last minute my opposite number at Lethebys, giving the document a final once-over, recognised me and I understand that he expressed more than a little surprise that his staff had allowed this upstart competitor to be on the front of his new, very expensive, brochure. What I'd have given to have been a fly-on-the-wall during that meeting – even better had the brochure been sent out to all their customers with Judy and me adorning it. Shame he noticed.

At the end of the following season, Letheby's contract came up for renewal and Lord March and the Goodwood management decided well in advance that they were going to put it out to tender. So in the spring of 1986 we, and the four other caterers considered capable of running the Goodwood operation, were invited along with the incumbent, to a presentation by Rod Fabricius of what was required. He stressed that what they were after was innovation, style and quality. He assured us that the contract would not necessarily go to the company with the deepest pockets.

So followed a detailed review for four months of Goodwood's entire summer season. I personally was present at the majority of racing days but at least one member of our team attended every single race meeting. We got to know the operation very well by the time we put together our written presentation, which was an extremely long and detailed document.

A month after this was submitted, the companies were called in one by one to be interviewed by the Goodwood board and to make our personal presentations. At P&G, we decided that four of us would make the presentation and we each had a specific section to talk about. I was to open the batting with an outline of our approach and then a final flourish at the end, Roger de Pilkyngton, our Sales & Marketing Director, would talk about packaging hospitality and developing sales using our database of clients as we had done so successfully for Rod Fabricius at Aintree. Bob Culver would give the financial overview and explain about the support from Job's Dairy and Neville Petts would describe the changes we would make to the public catering, speed of service and so on.

We agreed the amount of time that each of us would take, making sure that the whole thing lasted no more than thirty minutes. Scripts were written and presentations rehearsed in front of the Job's board together with questions and answers. We were ready to go.

At this stage, I have to explain about Neville. A short, powerfully-built man with a constantly worried expression, bustling with energy and forever pushing his glasses back onto his nose, I had, of course, known him, as a good friend and somewhat of a business mentor, for over twenty years by the time he joined us. I was, therefore, aware of his very considerable strengths as well as his eccentricities. He was a very professional manager who had had a succession of very high-powered jobs in the hotel industry.

However, besides sometimes going off at tangents, his considerable energy was not always totally focused on the right object. Neville was once brilliantly described (probably by me) as 'a bull who carries his own china shop around with him'.

An example was at Farnborough Air Show; visiting for the first time one of the chalets for which we were catering, Neville saw some chefs outside the back door smoking in full view of the public. He ordered them in to the kitchen, followed them in and berated them for several minutes about lack of discipline, how filthy the kitchen was and it was only when he stopped to take breath that one of the chefs was able to say,

"Excuse me sir, but who are you?"

Neville said grandly, "I'm Neville Petts and I'm a director of Payne & Gunter."

The chef said quietly, "Well Mr Petts, this is one of Ring & Brymer's chalets, P&G are next door."

Besides occasionally going off at half cock, he also had no sense of time and a very unconventional way with the English language. Whenever I think of Neville I am reminded of the wonderful story that Humphrey Lytletton, the jazz trumpeter and witty chairman of radio shows, tells about a time he was interviewed by a young journalist from a local paper who, having read about Humph's interests, said to him, "I understand that you are something of an orthonologist." Only to get the reply, "No I'm more of a word-botcher." Well that was Neville, a word-botcher; after four decades he is still liable to call me Ian.

So on the day itself, our presentation followed the one from the incumbents, in the Goodwood boardroom at the top of one of the two towers of Goodwood House, the outside of which is covered in flints chipped by Napoleonic prisoners of war. Everything went very well - although we were somewhat taken aback by the fact that Lord March was taking down everything we said in shorthand - seriously impressive - until that is, Neville's part in the proceedings.

Instead of the eight minutes we had rehearsed, he went off into a peroration of more than forty minutes on the public catering by which time Lord March, having already heard one set of presentations that day, was almost asleep. I was too far away from Neville to kick him so just sat and suffered along with everyone else.

However, since I was sitting opposite Lord March, and have a well-developed ability to read papers upside down, and with Neville going on endlessly, I had a good chance to review the schedule that Rod had prepared for his lordship detailing the salient points of all five tenders. To my horror I discovered that, not only were we offering less in concession royalties than two of our competitors, we were also way behind them in what capital we were prepared to invest.

There then followed a nail-biting three weeks before Rod phoned me late one afternoon to say that we had got the contract largely because of the innovation and our marketing ability. The great thing was that because I knew what the others had offered I also knew that Lord March and Rod had been as good as their word and not gone for the caterer offering the most money. This integrity repaid them handsomely as, having said as part of our presentation that we would increase the catering sales at Goodwood

Racecourse from the existing £550,000 to at least £800,000 in three years; we actually achieved sales of £1.2 million in the third year.

However, when Rod told me the good news that we had been appointed, he said that I had to keep quiet about it until he had phoned the others the following day. By one of those awkward coincidences, I had arranged for us to have dinner that evening with one of those competitors, Mark Butler, managing director of Ring & Brymer and his wife. Judy and I spent the whole evening trying to keep a smug smile off our faces.

Neville Petts was made the director responsible for Goodwood, which, since the racecourse was embarking on a programme to replace two buildings, was very helpful given his hotel development experience. It was, however, at Goodwood that Neville's eccentric ways really came to the fore.

In order to appreciate the situation fully, I must explain a little about one form of modern technology that progressively ruled our lives, the walkie-talkie. We all carried them and besides being an invaluable means of communication, it also was, for me, a fantastic way of keeping in touch with the entire operation as all messages went through the radio and I was, like GCHQ, listening to everything.

The machines were quite heavy to hold onto all the time so I had mine strapped to my belt with a curly wire connected to a speaking/listening microphone also clipped to the belt but which I could lift up when I wished to use the radio. This sometimes was embarrassing as on one day at a golf tournament when I was in the gents relieving myself alongside some golfers and a message suddenly started coming out of my belt. Instantly, the guy next to me turned to his friend and with a nod in my direction said, "I've heard people talking to their John Thomas before, but I've never heard one answer back."

It is easy to imagine that radios and Neville Petts were a volatile mix especially since, as I said, everyone heard everything. On one occasion he made a call on the walkie-talkie with the words, "Neville Potts to Simon Bailey" Simon picked this up instantly with a brisk, "Yes Mr Potts", which inevitably caused uproar among the 50 or so Payne & Gunter people in the radio loop.

However, the story that Neville will always be remembered for concerned a bomb scare during July Festival Meeting at Goodwood. The racecourse received a call saying that a bomb had been placed in the March Stand and would go off at 2.30. The stand was evacuated totally including the members of our staff who were the last to depart, leaving behind bar stocks, cash tills and so on. Because there had been experience of military radios triggering bombs, all our staff radios were ordered to be switched off.

Twenty minutes after the time the explosion was due to have happened, it was agreed that it had been a hoax. We wanted to get our staff in first, to safeguard the stock and cash, so the police nodded to Rod Fabricius who nodded to Neville, who switched on his radio and said into it, "All Payne & Gunter staff with their radios off please switch them on again".

He never lived that down and it was the subject of much hilarity at the end-of-festival staff dinner at a Chinese restaurant in Chichester as was the moment towards the end of the meal when Neville was served a rolled-up hot towel and tried to cut into it as if it was a spring roll. This was too much for one of the lads sitting opposite who laughed so much his chair fell over backwards. What we didn't know was that the curtain behind him concealed another room and he tumbled into this with the curtains closing instantly, causing him to disappear completely. It was a perfect antidote to the stresses of a week of catering at Goodwood.

During the winter, after the very successful first season, I met up with Rod Fabricius many times and on one occasion we had some laughs at the expense of our mutual friend John Hughes who, as Clerk of the Course of Chepstow Races, had been called to account at the Jockey Club because a race had been run round a shortened course as the dividing fences had been left in the wrong place. Inevitably the laughs of the racing world were based upon the assumption, quite reasonable, that Hughsie had been in a bar at the time.

The first day of the following season at Goodwood a similar thing happened and as luck would have it, it was for a Derby Trial, and as such a very high-profile race, made even more public because it was won by a very famous horse called Mtoto. After the race was over one of the jockeys said he thought that they had gone the wrong way. A video showed that he was right; the Goodwood ground staff had not moved a fence from the previous race and the horses missed out a loop of the track. Rod, as Clerk of the Course, was called before the Stewards at Goodwood and told that he would have to appear before the Jockey Club to answer the charge.

As Rod's wife Debbie was away, he stayed on at the racecourse after racing and he and I had a drink or two in his office. He was inconsolable, as he took his job and his reputation very seriously. I tried to temper his distress with comments such as, "Well Rod, when you employ people you have to carry the can for their mistakes".

"But I was hauled up before my own Stewards."

"We all have staff who do things wrong." "But my reputation." "No one will believe it was your fault." "But I've got to appear before the Jockey Club." "Well, they'll understand."

"But you don't understand Alan, what can I do? What can I say to them?"

By this point he was almost suicidal and I thought humour was probably the only answer. It could have ended a perfect friendship but I couldn't resist saying, "You could always ask them to take another seven fences into consideration."

For a second I thought he was about to hit me but he roared with laughter. In the end he didn't shoot himself, or even the guy who forgot to move the fences.

14

DEAD AS A DADA

Early in 1985, John Coomb and I found ourselves listening to a presentation by a senior representative of the Grace Corporation, an American chemical company who were proselytising about the use of a very modern cooking system known as 'sous-vide'. As an aside, I have to point out that I was under a certain amount of pressure about this because, not only was it very much the bee-of-the-moment in Neville Pett's bonnet, as the director in charge of production, but also my chairman, Simon Roberts, was absolutely in favour and had invested in a restaurant in Brighton totally based on the principle.

The man from Grace gave us a powerful presentation in our pantry with all the necessary visual aids and an acolyte who changed the slides and put the lights on and off. Grace-man pointed out that 'sous-vide' was a system of cooking food that had firstly been placed in plastic bags under pressure. Once cooked in this way the food could be stored at ambient temperature for up to three weeks, only needing regenerating in hot water before being opened and served.

The system, Grace-man pointed out, had been taken up by many of the top chefs in London and especially by the Roux Brothers who had opened a restaurant specialising in the service of 'sous-vide' menus.

Having patiently listened to all this, which was really no more than I had already heard several times from Neville and Simon, I decided to bowl my googly, "It's all very well for a restaurant but how about a banquet for seven hundred?"

I expressed my concerns and described my visions of twenty or so chefs faced with seven hundred sealed plastic bags, each filled with a cutlet of salmon, and another seven hundred, containing 'Breast of Duck Martine' and turning to one another saying, "I thought you were bringing the scissors."

Grace-man didn't laugh. He replied with a question, "Do you know the Roux Brothers?" When I replied that I did, he said, "They catered for a very special function at the Royal Albert Hall for 2500 guests using only 'sous-vide' food."

I nodded, showing interest but really wanting to encourage him to go into a story that I knew he was going to regret. Chuckling inwardly, I let him carry on extolling the wonderful benefits of 'sous-vide' and how successful the system had been when Albert and Michel Roux used it at this grand and stylish event. Neville jumped into the conversation, saying that John and I couldn't gainsay the new concept, on the grounds of quality or numbers, when it had received such a wonderful practical endorsement by such worthy and famous chefs.

John Coomb looked at me and gently raised an eyebrow as what he and I knew, and Grace-man and Neville did not, was that we both had been very much involved in that grand

and stylish event, some two years previously and before Neville joined us. It had been one of those occasions when I had allowed my curiosity, followed by my professional vanity, to lead me into a seriously hair-raising venture.

The saga started in April 1983, when Chris Stokes, our warehouse manager, received an enquiry from Monsieur Girardin, the general manager of Roux Restaurants, who wanted to hire 2500 covers of equipment, each one a complete place setting of cutlery, crockery and glass for three courses, for a night in June. Chris was very excited - I was intrigued.

Again I have to hark on about the changes that the last twenty-plus years have made. Nowadays anyone who has the chutzpah to convince themselves and a client that they can cater for a function for a couple of thousand people can do so. They merely have to phone up a number of companies who can deliver a myriad of support services from ready-made kitchens to entire teams of chefs, managers. Before these companies came along, there were probably only about half a dozen established firms who could be considered, including P&G.

However, the Roux Brothers at that time were the gastro-stars; they were the first restaurateurs in Britain to be awarded three Michelin stars and their 'Maurice Chevalier English' was heard regularly on television. Probably next in line from Escoffier and following the distinguished line from Carème and Soyer, they were creating the modern day phenomenon of the celebrity chef. I had known both of them for a dozen or so years at that time, having met them on a trip to Taittinger Champagne in Reims, and one of my earliest regrets is that I allowed a small event of ours make me turn down an invitation from Michel to the opening of the original Gavroche. We were not bosom buddies but met at various trade events and wine tastings and discussed such problems as dealing with the tronc, which is the way tips are distributed to staff, and the taxman.

I was fascinated how they, who ran three restaurants at that time, with a maximum of two hundred covers, were going to undertake an event for more than ten times that number. So I phoned Albert Roux and asked him precisely that question. He said, "I am sure that we will be able to recruit the necessary staff and my managers will be able to run the event."

I asked, "How are you going to be able to cook and serve such large numbers?"

He replied, "We are going to be using this wonderful new system called 'sous-vide', the food is put into plastic bags and cooked under pressure and then all we have to do is reheat just before service, which will make it very easy in the kitchen".

He said that he would be delighted to hear any thoughts I had on the matter and so I had a meeting with him and his brother at Gavroche, Albert's celebrated Mayfair restaurant, and, over a glass of Moët, he started to tell me all about it.

"We have agreed with the Design and Art Direction Association, DaDa, to do their dinner to celebrate their 21st birthday."

Michel chimed in, "Their members are the bosses of the advertising industry and are very good customers of Albert and mine". He continued, "It will be for 2500 people at the Royal Albert Hall, which my brother thinks was named in his honour. It will be a wonderful way of introducing this cooking system to a lot of people – we are opening a restaurant in the City which will only use 'sous-vide' and this dinner will be a showcase for us"

I found myself quite as charmed as everyone appears

to be when talking to these two great personalities. I actually started believing what they were telling me, but I wanted to know more, "How are you going to find two or three hundred staff for such an occasion, let alone the management?"

They said that they were working on this but had not as yet come up with a guaranteed supply and it was at this point I heard myself saying what turned out to be one of the dumbest offers I have ever made, "Why don't we look after the operational and front-of-house side for you and charge you for it?" Stupid, stupid, stupid.

One of our golden rules as caterers was that, since there are always so many circumstances that are outside our control, those that we could control, we did - ruthlessly.

The trouble was that the Roux Brothers and their restaurant manager, who had organised things up to this point, were not caterers, they were restaurateurs and they didn't know this golden rule, and they didn't know what hazards to look out for.

Inevitably all sorts of things had already been agreed with the client and the members of the association, between DaDa and the Albert Hall, and with Hedges and Butler, the wine merchant who would be organising the cellar on the evening.

Most of these arrangements were in stone by the time we became involved. Only slowly did this become apparent.

The first meeting at the Royal Albert Hall was an event in itself; it comprised M. Girardin, together with the Sous Chef from Roux Restaurants, the General Manager of the Albert Hall, six members of his staff, the resident catering manager, the security officer, four blue-chalk-stripe-suited, Jermyn-Street-shirted wine merchants from Hedges and Butler, the magnificently flamboyant Edward

Booth-Clibborn, Director of DaDa, (who, I swear, had an astrakhan-collared coat and a fedora), his two very pretty assistants, a couple of Dada council members - and me. A jolly, manageable team of twenty – just the way to run a party for two and a half thousand.

As the meeting progressed, it became clear that, at the beginning of the year, DaDa had appointed Roux Restaurants to do the catering and Hedges & Butler to supply the wines according to the menus and wine lists they had provided respectively. Publicity had been sent to the DaDa members several months before, soliciting their bookings and informing them what was to be available from these illustrious houses. Since then absolutely nothing had been done in preparation until the visit of M. Girardin to our warehouse, three weeks earlier. I was walking into quicksand.

One of the most useful weapons in a caterer's armoury is the ability to bend the client's requirements to what you know is possible, usually without them knowing. Over the years I had developed this ability to an art form, and so while the meeting was starting to fill me with a great deal of nervousness about what I had taken on, I was convinced that I would be able, by a mixture of sleight-of-hand and an element of arm-twisting, to make things considerably more manageable.

I was wrong. Celebrity chefs, upper-crust wine merchants and trade association directors used to getting their way against powerful members, do not take kindly to being told by an upstart caterer that their arrangements must be altered.

One of the main problems I foresaw was that a very elaborate wine list had been sent to the clients. "Was this practical?" I asked, "Couldn't we have a shorter, simpler list, perhaps four or five wines instead of twenty-five?"

DaDa-man assured me that his members were expecting a wine-list fully as elaborate as that they were used to browsing when they went to Gavroche or Michel's restaurant, The Waterside at Bray, which, of course, being advertising executives, they did quite regularly.

I then had the temerity to ask of the serried ranks of wine merchants, "Are you sure that you are going to be able to maintain this entire list through the evening – it is going to require a lot of wine".

This prompted a rejoinder with due hauteur from Hedges head honcho, " We are the Queen's Wine Merchant." This put the upstart caterer firmly in his place but not before I observed, "One of the Queen's Wine Merchants." I don't think this endeared me to him.

It soon became apparent that I was going to have to accept this job 'warts and all' or not at all. Having gathered all the information, I arranged to meet the DaDa and Roux people at the Albert Hall when I had had a chance to review what I needed to do.

There followed a frantic few days planning. It was going to be a lot more than merely hiring out 2500 covers of equipment. A great deal of work was going to be required in the preparation for the event as we were, to all intents and purposes, going to be running it, although our name was not to appear anywhere; Roux were to receive the kudos. At the time I was upset about this because of all the work we were going to do, without any credit, but at the end of the affair, this was going to be one decision I would look back on with delight.

Since the food was going to be prepared by this 'sous-vide' method in the basement of the Albert Hall and everything was to be plated, we had to design a system for transferring 2500 plates of each of three courses to all corners of the Albert Hall on seven floors.

This required purchasing 7500 plate rings, 500 large plastic trays, and having 100 tall trolleys made to specification in order to transport the plated meals around the building.

Because the door to the Albert Hall that we were going to use for access to our distribution room was at the top of a short flight of stairs from the pavement, we had to manufacture a massive 'drawbridge' to go across the pavement from the back of our vans to the doorway, some twelve feet.

We would need to organise 260 waiting staff for food service, 50 wine staff, 40 porters and a management team of 30. All of these staff needed to be instructed and trained for the night. And we had five weeks.

After all this information had been put together, I had a very serious discussion with my team about whether, given the time at our disposal and the difficulties in the arrangements we had inherited, we could actually do it. This was where our professional vanity kicked in. In spite of the fact that this was the beginning of our summer season and so we were already becoming very busy, we decided that we could and should do it, if DaDa and Roux agreed to the costs involved.

We also realised that with only five weeks to go, if we decided not to undertake it, there was no way it was going to happen - and we would get the blame – damned if we do, damned if we don't - better to do it.

So after a considerable amount of discussion and agreement on the costs, we set about the mammoth task of preparing for the evening and, following countless preparatory visits to the Albert Hall going through the details of our plans with their catering and security staff, we arrived there the day before the event with our vanloads of equipment. The nightmare began.

The Royal Albert Hall was built from the proceeds of Prince Albert's Great Exhibition of 1851, as was virtually the whole of the museum area of South Kensington. It was always intended to be a general-purpose building and had the aura of being an architectural 'jack-of-all-trades, master-of-none'. Certainly it was not designed to stage vast dinner parties and had no infrastructure for this purpose. Just prior to the millennium, many millions of pounds were spent dramatically improving the access and means of operating the building, but at the time of the DaDa evening, in the early eighties, we were still using the original hundred-and-thirty-year-old facilities.

The dinner utilised the whole of the building. In front of the stage is the Arena where, on the last night of the Proms, the Promenaders can be seen on television swaying and singing patriotically. On this occasion it was covered with twenty large tables for the major companies of the industry, such as Saatchi & Saatchi, McCann's and Collett-Dickinson, as well as DaDa themselves.

The next area was the Stalls, where the ten rows of tiered seating was transformed into five rows of dining tables, only facing one way, of course. This was done by covering every other row of seats with 'mountain-goat' trestle tables with their front legs longer than the back. These were supplied on hire and set up by Benjamin Edgington, the Queen's marquee contractor, having been made by them originally in the 1890's - a time when Edgingtons were Queen Victoria's Ball Furnishers.

Behind the Stalls were the 30 Loggia Boxes, above which were 36 Grand Tier Boxes, on top of which were 72

Upper Tier Boxes. The tiered balcony seating above these was not used so the final area operated for the dinner was the flat Gallery at the very top of the building where six hundred places were set up. Very, very remote, it was obviously for the also-rans of the advertising industry. This section was managed by Bernard Clements, the catering manager we inherited from Gunters, and, because of its lack of access to anything, least of all service, he referred to it as 'Spike Island'– where he was marooned for the night.

The Roux Brother's kitchen was set up in a large room in the basement. Two wine dispenses were set up, one on the ground floor level, serving the Arena, Stalls and Loggia Boxes and the second on the Grand Tier level serving all the upper floors – this was therefore five floors away from Spike Island.

We had a largish room, maybe forty foot square, on the ground floor level as a distribution room with access to the outside, where our 'drawbridge' was set up connecting it to our vans. After the equipment had been circulated to all areas, this room was to be the 'clear-area' during the dinner where all the dirty china and cutlery was to be packed and loaded for return to our depot.

There was only one goods lift in the building that could be used to go down into the kitchen or up as far as Spike Island; it was therefore going to be critical that its use be carefully scheduled with Chris Stokes, our warehouse manager in total charge. First the distribution of the incoming equipment, then the wine stock into the dispenses, then the food into the kitchen, the wine orders distributed to service points, the first course from the kitchen to service points, the second course to the service points while the first is being eaten, the dirty plates from the first course to the clear, the dessert to the service points, the dirty plates from the main course to the clear, the coffee and petit fours in, the dirty dessert plates out – all the while topping up the wine orders – then clear the coffee cups and finally clear off the tables – and to think one considered the job of beach-master in Normandy on D-Day as difficult! It didn't quite work that way.

When we arrived on the morning of the day before the dinner, it was delightfully tranquil, we had the place to ourselves distributing our equipment, until the arrival of the wine merchant's vans and their drivers and porters who dispatched the wine in two separate lots to the dispenses. This did somewhat concern me as it appeared that the wine had not been broken down into individual orders which would mean that a great deal of work needed to be done on site. It concerned me even more when no one else from the wine merchants arrived that day to do any sorting out; everything was going to be left till the day itself. I was even more concerned with the team who arrived the following morning.

The West End wine trade at that time was only just escaping from the affliction that had cursed it for the previous two centuries; the need to employ the fourth son of all their landed clients; the ones who couldn't get a commission or join the ministry of the Church of England or settle the Empire: the sort who, several years later, were drawn so tellingly by Harry Enfield as Tim-nice-but-dim. Obviously we had the last battalion.

What concerned me was that the wine merchants had received beforehand 382 individual wine orders from 382 individual, highly vocal and very demanding hosts. It now appeared that these were to be made up by a team who, on an escapade at the time of a previous generation, you would have preferred to have left back at the Château with General Haig rather than have alongside you in the trenches. Although this was no part of our brief, I immediately phoned back to our HQ in Chiswick and ordered our West Indian cellar master, Vic Kirnan, and his team to come down immediately to give some structure to the proceedings.

Even he, with all his experience, found it difficult because, since the orders had not been collated properly, either they were, as I hinted at before the event, running out of certain wines and needing to make substitutions or the appropriate wine was in the other dispense. It was a plumber's muddle of the first order.

The knock-on from this was considerable; firstly the wine distribution over ran its allocated time in the lift with dramatic effect. This was not as great a problem as we encountered when the 382 hosts arrived and many of them discovered that they did not have the wine they had ordered, or indeed, had none at all.

To gauge the seriousness of the problem it is necessary to realize the nature of the evening and the membership of DaDa. This was Ego Evening. We had relatively little problem with the big boys – they had nothing to prove. This was, however, the evening where the smaller operators, the new ad agencies, just started and on their way up, considered themselves on equal footing with their illustrious peers. This was probably the only time they would be served by the Roux Brothers and they had been led to expect the best for their guests, who they were trying to impress, and who were, after all, the only ones that mattered in the Albert Hall that night.

The testosterone-charged reaction to the wrong wine had to be dealt with by our poor hapless wine service staff, like dear old Iris Bloomfield, who, to correct the position, had to make her way back to the wine dispense, the other side of the building, trying to ignore the bunions, (her lifetime-award from the catering industry), and then return with a further substitution, not knowing whether that was going to be acceptable or not to the, by now, totally unreasonable host. This story was repeated by all our fifty wine staff who spent the entire evening to and fro in total frustration trying to correct a basic flaw that wasn't at all their fault – we're back to the trenches.

Meanwhile, in the kitchen, strange things were happening.

We were very strictly not allowed in the kitchen that had been created in the basement but occasionally we caught a glimpse of a chef as he came out for a cigarette. Because of the 'sous-vide' system they were all gowned in green with green surgical masks and caps, looking like extras from ER. This was all somehow to do with the new system, although it was never made clear whether they were afraid of catching something from the food or vice versa. If it was vice versa, I could not see how we were going to protect the food on its hazardous journey once it left the sterile kitchen. Heaven knows what the customers might pass on to the food once it was out in the open.

Twenty of our new trolleys, each containing 75 plates of the first course, were to be taken to the clear area on the ground floor level to supply the 1500 people in the Arena, Stalls and Loggia Boxes.

We had agreed with the Albert Hall management that, once the guests had passed through to their seats, these trolleys would be rolled out along the corridor running round the building behind the Loggia Boxes and be based, with accompanying porters, alongside each of the six staircases coming up through the Stalls from the Arena so as to act as collection and delivery point with each trolley supplying 8 waiting staff. The other thirteen trolleys for the first course would be similarly dispatched to staging points on the other floors.

The plan was that, after the first course had been consumed, the dirty plates would be returned to the trolleys and wheeled back tidily to the 'clear-area' to be packed and loaded on the vans. Meanwhile at the staging post the trolleys with the next course would have arrived.

To coordinate the service, we had our urbane City Director, John Coomb, standing on the stage, looking just like a maestro conducting an orchestra, with his dinner jacket and his flowing grey hair. We had rehearsed all afternoon that when he heard by radio that something was ready to be served or cleared, he was to raise his white-gloved hands, at which a sound signal went plinc-plonc-plinc around the Albert Hall. On hearing this sound, 260 waiting staff would move as one to serve the first course. There would be a similar performance to clear the first course plates away, and so on through the meal.

I have often likened John's ability to coordinate his service to the films of Busby Berkeley, with their magnificently synchronised 1930s beauties and gorgeous art deco sets, and this time John meant to exceed even his own normally high standards of theatricality. In the event a more apposite film simile would have been Keystone Kops.

The first course was to be a couple of softly poached eggs, which had, according to Michel Roux's pre-publicity, been "gathered at dawn the previous day from 'appy French 'ens".

When the two thousand odd guests arrived to be greeted with their reception glass of Champagne and the 382 hosts did battle with the wine staff, the eggs from the

’appy ’ens had just about made it to the Loggia level in the ‘clear-area’ but we were still waiting for them on the other floors, it was taking longer than the Roux Brothers had anticipated. It was at this point that catastrophe arrived, in human form.

The Albert Hall catering manager introduced me to the fire officer who told me quite categorically and without any preamble that there was no way that we could bring our trolleys into the corridor behind the Loggia Boxes. Protestations by me that the corridor was twenty foot wide, and why tell me at this time when the Albert Hall had known about the plan for weeks, were to no avail and failure to comply would result in the whole evening being stopped – not a pleasant prospect – but neither was changing the entire service plan for two thirds of the event while the guests were having their pre-dinner drinks. I could tell by the sheepish look of the catering manager that this was not a total surprise to him.

Whether or not this was a whim of the fire officer’s vanity, I had to comply, no point losing my temper; I had a hundred and fifty staff to give new instructions to. However I did lose my temper about the unfairness of his ultimatum when I discovered at the end of the evening that the fire escape for Spike Island, where six hundred people had been dining in the roof of the building, had remained chained and padlocked all night; this he had quite shamefully overlooked, while concentrating his attention on my inoffensive and harmless trolleys.

Nothing for it, the ‘clear area’ instantly had to become the servery for all 150 waiting staff. This meant that, because the corridor did not go across the back of the stage area, those with stations on the other side of the Albert Hall had to walk around three-quarters of the building with each load of three or four plates, up to four trips for each course, assuming that they remembered which staircase they had to go down, since they all looked alike.

It was inevitable that this caused a considerable delay in serving the first course and made somewhat of a mockery of the radio-controlled John Coomb on the stage with his white gloves and his plinc-plonc-plinc. I hoped that we could take advantage of this delay, because I hadn't at this stage seen any of the main course arrive from the basement; discreet word came through that the much-vaunted 'sous-vide' system was not living up to its promises.

Eventually, still with no sign of the main course, we couldn't delay the clearing of the first course, which I knew was going to mean pandemonium in the 'clear area' with a hundred and fifty waiting staff arriving, each with ten dirty plates, into an area that was already congested with trolleys and packing cases. Almost as soon as I had sent the waiting staff out to clear the first course plates from the tables, I was horrified by the immediate arrival from the kitchen of the twenty trolleys with the main course, just when we didn't need them.

Five minutes later, John Coomb came from his place at the front of the Hall to find out why the delay and where the main course was. The scene he was faced with was out of a Hieronymus Bosch painting – a room the size of a classroom, which was accommodating in order: a dozen and a half porters, the piles of empty equipment boxes ready to take everything away, the twenty trolleys from the first course, one thousand five hundred dirty plates, three thousand pieces of dirty cutlery, one hundred and fifty waiting staff, twenty trolleys filled with the main course ready to go out - if anyone could get to them – and me.

The walkie-talkie radios, meanwhile, were continuing

to tell of ongoing skirmishes in the wine order war.

John looked at me and said, for the first time in his career, "Shall we fuck off home?" Also for the first time in my career, I was inclined to say yes, when, magically, we were saved.

The double-doors at the far end of the room burst open and in marched the Band of the Brigade of Ghurkhas – playing Blaze Away. Apparently they were to be the inter-course entertainment and had come in the wrong door.

The unexpectedness of their arrival and the comedy of watching them thread their way through the mêlée, while still playing, was enough to create absolutely hysterical and helpless mirth among a group of previously suicidal caterers.

In their magnificent uniforms they made an impressive, if rather disjointed, sight as they used their famed dexterity-in-combat to avoid the remains of the 'appy 'eggs. I was grateful that their continued use of their musical instruments meant that it was impossible for them to extricate their kukris and hack their way through the scrum, as they probably would have liked to do.

Seeing them battle their way through the serried ranks of Payne & Gunter waitresses, one didn't wonder about their dreaded reputation as jungle warriors (the Ghurkhas, that is, not our waitresses – although they too are fearsome foes).

While we never caught up with the plot, after that we did at least keep it in sight, and were able to take with equanimity the radio crossfire, half an hour later, when John Coomb demanded, "Where are the petit fours for the Arena and Stalls?" Only to be met with the plaintiff cry from Bernard Clements, up in Spike Island, saying, "Sod your petit fours, I haven't had my main course yet!"

We'd forgotten all about him and his six hundred.

It was at about this time that the Director of DaDa,

Edward Booth-Clibborn, like the Victorian actor-manager that he should have been, stood up before the massed ranks of his members and their guests and introduced the famous chefs responsible for the magnificent dinner. The kindest thing to say was that Albert and Michel Roux came on stage to a less ecstatic reception than they expected.

The rest of the evening was enormously hard work but an anti-climax. After the guests left, we spent the entire night clearing up and packing everything and when I arrived home the polling-station across from my house was just opening – apparently while we were sorting out this event an entire general election campaign had been and gone – at least there was one thing in DaDa's favour.

There inevitably followed a dreadful period of recrimination as success has many fathers but failure is always an orphan. We also wanted to get paid and that took a very, very long time. This whole dispiriting post mortem period was singularly brightened up by reading the letters of complaint from the members of DaDa, who, given the nature of their trade, had a nice line in controlled vitriol.

My favourite was from the boss of Collett Dickinson who said that, "had Albert Roux a vestige of honour he would have committed hara-kiri with his Sabatier".

It is now possible to see why it was that I so enjoyed the discomfiture of Neville and Grace-man when I related the story to them and suggested that perhaps the event should not be used by Grace-man as a sales aid. Needless to say, Payne & Gunter never did succumb to the joys of 'sous-vide'.

15

STAFF OF LIFE

When friends or customers hear about the chaos of my many near-escapes, like the DaDa evening, they usually ask me something on the lines of, "How on earth did you cope with living on such a knife edge?" Such a question usually elicits the droll response from me, "If I allowed myself to worry about all the things that could possibly go wrong, I wouldn't have a hair left on my head." Typically, this prompts an instinctive laugh, which turns into a more nervous chuckle as they glance at my very bald head.

The uncertainty created by my retort is, of course, entirely intentional, as already explained while I had an ability to create a 'jolly shell' this doesn't mean that I was immune to the pressures.

The main reason throughout my career that I was often never more than a moment away from sheer blind panic (even at times of almost sublime excitement, with an event going according to plan and customers eating wonderful food in exquisite surroundings) has been because of one word - staff.

As I explained to Matt Busby, during that lunch at the Open Championship at Muirfield, given the size of our events my success or failure was, from very early days, always going to be down to the staff I employed. Fortunately, over the past four decades, the vast majority of those I have worked with, both permanent and casual, have been an unmitigated joy. Well, maybe sometimes the joy was mitigated. A small percentage has been slightly unreliable and a tiny minority has been, inevitably, plain bloody impossible.

The first difficulty was, of course, knowing t'other from which. There was also the small problem, even with those who were wonderful, reliable, charming and competent when they were on site, of whether they would get on site and, if so, when.

As our events got bigger and bigger, the waiting staff had to be recruited from further and further afield, their journeys became longer and longer and in some instances we only expected them to arrive in time for lunch service, and so two-thirds of the day we were paying for was spent in a coach.

This had the double effect of the staff getting caught up in the considerable traffic generated by the events themselves, instead of arriving as the rest of us did before the crowds and, when they did arrive, we had no prior warning of who exactly was turning up and no option but to put them straight to work – somewhat risky, at the very least.

The largest part of our casual recruitment was for waiting staff. These had the added hazard that they were our visible face and therefore represented us to our customers. By the 1980s, when we were catering at many locations on a regular basis, we had extremely reliable teams of 'permanent casuals' based around both the City of London and at

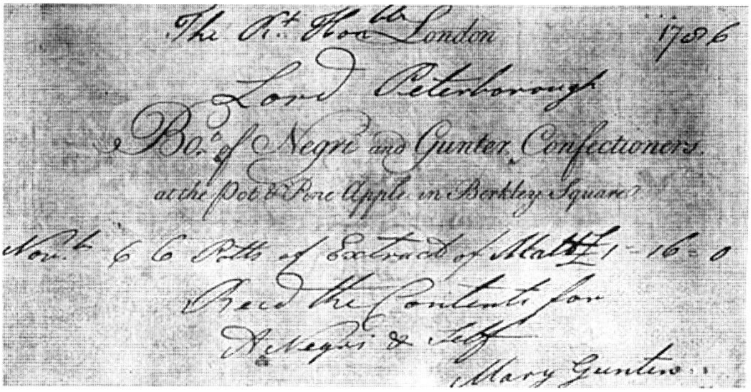

The Rt Honble London 1786

Lord Peterborough

Bot of Negri and Gunter, Confectioners,

at the Pot & Pine Apple in Berkley Square

Novr 6 6 Potts of Extract of Malt £1-16-0

Recd the Contents for

D. Negri & Self

Mary Gunter.

Invoice dated 1786 Negri & Gunter

MINISTÈRE DE LA MAISON DE L'EMPEREUR

Brevet de Fournisseur de l'Empereur

N° d'Ordre 122

L'Empereur Napoléon

Sur les témoignages avantageux qui lui ont été rendus de la moralité de Messieurs Gunter et Cie à Londres, Berkeley Square, [illegible] ainsi que de la réputation distinguée qu'ils se sont acquise dans leur profession, désirant leur donner une marque particulière de sa bienveillance et de sa protection, Nous a ordonné de leur accorder le titre de Fournisseurs de l'Empereur.

En foi de quoi nous avons délivré le présent Brevet aux Sieurs Gunter et Cie.

Donné à Paris le [illegible] Mil huit cent [illegible]

Le Ministre d'Etat & de la Maison de l'Empereur

[illegible]

Royal Warrant from Empereur Napoleon dated 1855

William Gunter (taken from Gunter's Confectioner's Oracle 1831)

Advert for Mayfair Catering 1932

Battle of Britain Reception Guildhall 1980

Battle of Britain Buffet Guildhall 1980

Restaurant in the Sky Aintree 1984

The Two Alberts, Roux and Hall

Simon Robert's 50th Party Toy Museum

Emma Payne's 21st on restored Temple Island

Edwardian Banquet at Tallow Chandlers' Hall 1985

Replica of Gunters' Cake for Queen Victoria's Diamond Jubilee

'William Gunter' and Roger de Pilkyngton off to Chicago 1990

Dinner for Her Majesty at Hampton Court

HENLEY
ROYAL REGATTA
Banquet to mark the occasion of the
150th Anniversary
of the Foundation of
Henley Royal Regatta
for the benefit of the
Temple Island Restoration Fund
Sponsored and Presented by Payne & Gunter
Menu
As served by Messrs Gunter to Her Imperial Majesty Queen Victoria
at her various palaces and courts

Veuve Clicquot NV	Canapés
Rainwater Barbieto Madiera	Potage "Carlton House"
Hochheimer Königen Victoriaberg 1983	Filet de Truite à la Valencienne
Château Gazin 1981	Pintade
	Sorbet au "The Gunter"
Volnay Louis Latour 1982	Baron d'Agneau
Veuve Clicquot Rich England	Macedoine des Fruits au Clicquot
Taylors 10 year Old Tawny	Petits Fours
J de Malliac VSOP Armagnac	Café

Pageant for the Banquet
William Gunter, Esquire
By appointment confectioner to Her Majesty Queen Victoria
Toasts
Her Majesty The Queen
The Lord Mayor, Corporation of London and The Sheriffs
Music
Harp & Clarinet Duo with Danielle Perrett and Graham Scrivener
The Stuart James String Quartet
The Herald Trumpets of Thames Fanfare Brass
Speeches
The Regatta
Alan Payne, Managing Director, Payne & Gunter
The Guests
Peter Coni, OBE, QC, Chairman, Henley Royal Regatta
Response
Peter Yarranton, Chairman, London Council for Sport & Recreation
Vice-President, RFU.
Beating Retreat
The Band of The Irish Guards
Guildhall
Thursday, 18th May, 1989
Payne & Gunter

Menu for Henley Royal Regatta 150th Banquet 1989

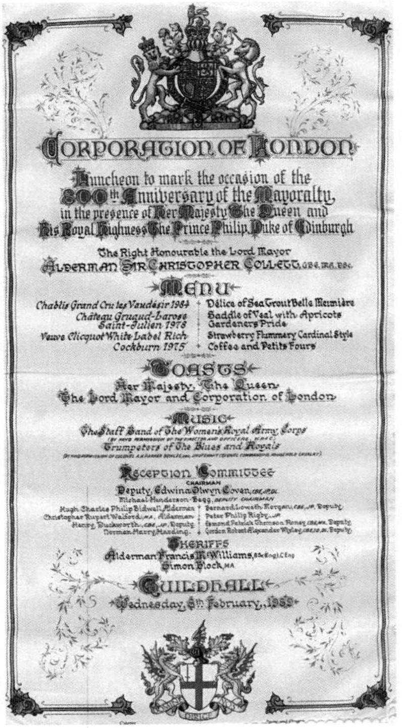
Corporation of London
Luncheon to mark the occasion of the
800th Anniversary of the Mayoralty,
in the presence of Her Majesty The Queen and
His Royal Highness The Prince Philip, Duke of Edinburgh
The Right Honourable the Lord Mayor
Alderman Sir Christopher Collett, GBE, MA, DSc
Menu

Chablis Grand Cru les Vaudésir 1984	Délice of Sea Trout Belle Meunière
Château Gruaud-Larose Saint-Julien 1978	Saddle of Veal with Apricots Gardeners Pride
Veuve Clicquot White Label Rich	Strawberry Flummery Cardinal Style
Cockburn 1975	Coffee and Petits Fours

Toasts
Her Majesty The Queen
The Lord Mayor and Corporation of London
Music
The Staff Band of The Women's Royal Army Corps
Trumpeters of The Blues and Royals
Reception Committee
Chairman
Deputy, Edwina Olwyn Coven, CBE, JP, DL
Michael Henderson-Begg, Deputy Chairman

Hugh Charles Philip Bidwell, Alderman	Bernard Lowath Morgan, CBE, JP, Deputy
Christopher Rupert Walford, MA, Alderman	Peter Philip Rigby, JP
Henry Duckworth, CBE, JP, Deputy	Esmond Patrick Thomson Roney, CBE, MA, Deputy
Norman Harry Harding	Gordon Robert Alexander Wixley, CBE, TD, DL, Deputy

Sheriffs
Alderman Francis McWilliams, BSc(Eng), CEng
Simon Block, MA
Guildhall
Wednesday, 8th February, 1989

Menu for 800th Anniversary Banquet Guildhall 1988

Twickenham and, give or take the occasional cock-up, this worked pretty well.

The problems when they arose nearly always concerned the larger events; a major banquet at Guildhall requiring eighty or ninety waiting staff – for one evening; a major golf tournament in Scotland, Wales or anywhere in England requiring up to four or five hundred – for four days; a match at Twickenham requiring, by the end of my time there, up to two thousand – for one day!

So where from? We used some commercial recruitment agencies, but they were a very mixed blessing. Catering colleges were a source of both students and lecturers (as chefs or managers) to us, and we were a source of on-the-job training and very useful extra income to them.

However, by far the largest source was the network of 'Queen Bees' that we had established around the country over a period of fifty years. A magical series of Sarah Gamps (perhaps a bit unkind) who, while starting as waitresses, gradually took upon themselves to act as brokers of work for their local friends. Progressively, in return for a modest booking fee, they became a valuable resource for us, and our competitors, by organising a known quantity of staff and the travel arrangements to get them on site.

Our specific problem was that our major activity, golf tournaments, did not always take place at the same course each year – in fact the biggest of the lot, the Open Championship, moved round a regular tour of six golf courses in England and Scotland. This meant that we sometimes didn't use all of the 'Queen Bees' each and every year. On returning to their patch, we would contact them and a conversation would take place on the lines of, "The Open is back in Scotland this year – at St Andrews, Lizzie, are you alright for the same girls as before?"

"Hello hen, good to hear you. Absolutely! Fifty, as before?" Confusingly, Lizzie called everyone, male or female, 'Hen'.

In reality, the conversation took a wee bit longer than that. But, we were then secure in the knowledge that fifty of our required four hundred were coming from Lizzie. Known girls who knew us and knew what to do.

However, after the conversation would come Lizzie's stocktaking as, since Turnbury two years before, poor Ethel had a relapse, Doris has gone off with a ne'er-do-well, Gladys is doing silver service in the dining room in the sky and Beryl has never been the same since her hysterectomy, and so on.

However, on the bright side, wee Susan and her friend Rachel could get some time off school, and certainly can find black dresses; and hasn't Myrtle got a wee pal who's done some work at Butlins? And so on. As a consequence, at St Andrews fifty might well turn up, but sure as hell not the fifty we expected and knew.

Added to the doubts about supply, was the peril of the journey, which in days before mobile phones meant hours and hours of absolute uncertainty for the wretched caterer awaiting his staff on site. The problems could be man-made - a traffic jam or Pepe rolling over the minibus - or the weather.

One of our stalwarts was June Andrews who, with her Birmingham Girls, came to most of our events, including Twickenham for match days. My most enduring memory of June is when I was woken up by the phone ringing at four in the morning and hearing her Birmingham tone saying, "It's thick snow oop 'ere. Do you still want uz to cum?" A look out of the window told me that it was also, totally un-forecast, thick snow in Richmond, "Yes, June, the Match will still take place. Give me a call when you stop for a pee on the motorway."

'The Match' was the Varsity Match at Twickenham, our first one and exactly a year since Ring & Brymer had had to show us round before reluctantly handing over the contract. In the meantime we had successfully installed the Sweetheart beer pouring machine and a great number of other innovations, and now a new season at the RFU was about to start. Or maybe not.

Immediately, I got dressed and slithered my way through the muffled white streets of virgin snow to British Grove, knowing that I was going to have a busy few hours coordinating everything going out to all our City lunches as well as Twickenham, because the snow was bound to slow everything down to snail's pace.

Eventually I was able to get away from Chiswick at ten, and an hour later had arrived at the stadium, just a few miles away, where I reported to John Coomb, who was in charge that day. At that time the Varsity match was a very minor affair, although important to the RFU, many of whose members were Oxbridge alumni, with only about 12000 visitors and lunches being served only to the President's Room, the Committee and the Stewards.

It was not considered an event for corporate hospitality until we started to force our customers to entertain at the Varsity by using the leverage of places at the Internationals. Once they discovered what a great occasion it was, and how unique for midweek December, by the late eighties the match truly became the City's Christmas Party, and as big a draw as the Internationals themselves.

On the first Tuesday of December in 1981, I had not anticipated having to work at the Match, I was due to be there merely as a guest of the RFU at the opening of ERIC, the English Rugby Internationals Club. However, the snow had changed all that and one look at John Coomb's harassed face told me there was a job to do.

"OK, Coomb, where do you want me?" "Stewards' Restaurant, two hundred and twenty lunches" – no preamble. "What staff have I got? When does lunch service start?" "One trainee chef and Miss Knight and," he looked at his watch, "lunch started ten minutes ago."

I skated across to the Stewards to discover that my two hundred and twenty customers were there before me. I had to force my way through the very bulky mass of old rugger players, made even bulkier by their weather-inspired clothing, to get inside and get started.

When I eventually got into the restaurant, I found the young chef well up to the mark and efficiently preparing everything, but not yet anywhere near ready. All the equipment was still in the boxes in which it arrived from British Grove the day before except for a few soupspoons, which were being taken out and slowly and methodically polished by Miss Knight.

Miss Knight was a rather simple soul who turned up at some of our other larger events where her limitations where lost among the throng of other staff. Here, she was all there was to help me to serve a couple of hundred very hungry men. Given the fact that the Stewards' only reward for their services at Twickenham would be a ticket for a future match and the lunch they would receive, the lunch was pretty important.

Summing up the situation, it appeared pretty dire. Me and a dizzy waitress and a young lad in the kitchen, no one to run the bar, which should have been open by this time, and a lot of very thirsty, hungry and soon-to-be-very-angry large men. The Sweetheart Machine came to mind – the first day of operation for the second unit. I radioed across to the West Bar, where the machine was running, and told them to send over two hundred and twenty pints of beer immediately.

I then jumped up onto the bar counter and called out to the mass of disgruntled Stewards, "OK, chaps, we've got a problem. We can't open the bar because there is no one to run it." Muttered grumbles. "I have arranged for a pint of beer for each of you to come over in a couple of minutes, for which there will be no charge." Muted cheers. "Drink it slowly, because nothing else is ready yet. In ten minutes - or so, if you queue up tidily, I will serve you with a bowl of soup." Happier mutters. "If you bring your plate back, I will then give you a main course and later on, hopefully some pudding." Loud cheers – not for me but the beer arriving.

So that was how it worked. It was great fun and did my reputation at Twickenham no harm at all. The biggest cheer came right at the very end of lunch when June Andrews and the Brummy girls arrived, only to be told by the Chief Steward that they weren't needed any more as the guvnor was going to take over the Stewards in future.

The late-arriving Birmingham Coach became a standing joke. The most poignant repeat of which came at the memorial service we held for Gus's old catering director, Alan Cox, who had died suddenly only two years after retiring to live in Spain. The service, in the lovely Regency church of St Peter's, Hammersmith, not a couple of hundred yards from British Grove, was made truly memorable when halfway though, the doors creaked open and there issued the sound of twenty chattering waitresses, quite obviously from Birmingham, taking their belated places at the back. I could almost hear Coxie's rueful laughter.

Waking to dramatic and unexpected weather conditions has always been an occupational norm in an industry where climatic conditions are critical, so it took something special to really throw me. One such was the infamous non-hurricane of Michael Fish in 1987.

I in fact slept through the storm and woke up early, as normal, got ready for work and went out to my car. The scene that confronted me was one of total confusion. Very fortunately the massive sycamore tree in front of our house had shed its leaves a few weeks before, unlike most of the trees that were blown down, otherwise our house would have been completely destroyed.

As I tried to manoeuvre myself round fallen trees on my way to British Grove, one quirky thought kept entering my mind, something from Macbeth about Birnam Wood moving to Dunsinane.

At the time of the storm we were making the final preparations for the last tournament of the year, the World Match Play Championship at Wentworth Golf Club. Since this is in the middle of a wood and only a few miles away, I was not hopeful and sure enough virtually all the tentage was blown away.

This prompted one of the coolest remarks ever from our telephonist when an insurance company client whose business was drastically affected by the hurricane, phoned to cancel his pavilion at the tournament, "Certainly, sir, just as soon as we find it."

If waitresses were the most numerous of our temporary staff, but basically reasonably passive, the chefs were an altogether different kettle of fish. They came in ones and twos rather than fifties and largely as part of the freemasonry of chefs, by and large the network of whoever was our head chef at the time.

As I discovered when I was a young commis waiter at the Belfry Club, chefs are expected to be temperamental, in part purely because it is expected that they should be temperamental. There is also the fact they are often under severe time constraints and the self imposed pressure that come from a desire to serve a perfect meal.

They are also, it must be said, a randy lot. This, coupled with the opportunities that came from staying away from home, with a ready supply of sometimes quite attractive waiting staff, meant that life in outdoor catering was quite as steamy as any TV soap.

During the late eighties, while we were doing a golf tournament at Moor Park, one of the tabloids published a story about a leading PGA Tour golfer having an affair with a sixteen-year-old virgin who was working as a waitress at the event. The Payne & Gunter chefs were livid. Not that they were concerned about the morality of a golf star taking advantage of a young girl, merely that their manhood had been called into question over there being a virgin still working at Payne & Gunter. How on earth had that happened? They needn't have worried as the following week's tabloid reported that George Best had already scored with the said waitress, before she worked for us.

The food we served during Gus's lifetime, indeed until John Coomb took us into the City, was, although streets ahead of our outdoor catering rivals, still pretty simple and mundane by first-class restaurant and hotel standards. For a non-chef John had considerable culinary knowledge, which kept chefs on their toes. He also had very high aspirations for the menus he wanted to serve, and one of the reasons for his desire to join us was the standardisation that Trust House Forte put upon his small part of their empire.

One of the things that he was adamant about was

that meat, once carved, never, ever, went back into the hot cupboard before being served. This ensured that the meat was always pink rather than brown or grey and never had a dried skin of gravy on top, but it also meant that the chefs were always under pressure. For example when we catered for a banquet for seven hundred at Guildhall, during the brief time that the fish course was being eaten, nineteen chefs were carving the meat.

Needless to say the standards that John Coomb required, together with the more extravagant menus he promised his clients, put an ever-greater pressure on the kitchen brigade, a group renowned for their volatile temperament. In reality they handled the pressure magnificently but were never far away from an explosion not least in the run up to a major banquet.

The most memorable occasion where their equanimity was severely tested was at one of the most splendid and high-profiled events I ever attended. The Corporation of the City of London were hosting a State Luncheon to celebrate the 40th anniversary of the Queen's Accession, with Her Majesty, of course, as guest. To be held at Guildhall at the end of November 1992 it was, inevitably, the most hotly contested prize among the usual suspects of caterers.

John was at his zenith at that time and offered the committee a menu to remember; a Delice of Turbot followed by individual Roast Partridge, filled with Guinea Fowl and finally a Ruby Soufflé – not bad for six hundred guests. While it might have been a menu to soften the hardest of committee members, it did without doubt serve as a monumental challenge to our head chef, John Slight, and his brigade.

Sadly, what should have been a truly joyous occasion turned out to be a very sombre affair and was remembered for all the wrong reasons. A couple of nights before, a large part of Windsor Castle was engulfed in a terrible fire and Her Majesty standing in the November air observing the rescuing of irreplaceable paintings and treasures had, not surprisingly, been left with a dreadful cold and came to the lunch in a less than sunny mood.

After she arrived, it was decided that she would make her speech before lunch and not afterwards as anticipated. This meant that there would be a delay of twenty minutes. A breakdown in communication meant that we were not told until just before we were about to serve. It is safe to say that the chefs when they heard and with such a complicated menu to be put on the back burner were also in a less than sunny mood.

The speech the Queen made before the lunch became one of the most famous of her reign as it was the one where she referred to her Annus Horribilus; a year of highly publicised matrimonial problems for all her children and culminating in a fire at her beloved home.

The mundane truth for me was that I didn't see or hear the Queen deliver such a historic speech as, together with John Coomb, I was in the kitchen trying to calm down our splenetic and seriously frustrated chefs and stop them adding to the woe of the occasion by assassinating Her Majesty with their carving knives.

Since at any time we had only one head chef and the quality of the incumbent continually improved over the decade and a half after John Coomb joined us, it did not take long for his standards to cross into the events side of the business. This meant that all our casual chefs had to move up several notches along with the menus we were offering our clients.

At the major events we might have six or eight individual

kitchens, each of which might serve several hundred meals and so required a kitchen team, with a highly qualified chef in charge. These came from the Services, Army or RAF, or were lecturers at catering colleges, and some were restaurant chefs who took time off from their 'day-jobs' to earn some cash and have some fun.

One of the most colourful of these – in many senses – was Ainsley Harriott, who worked at many events with us. He almost invariably arrived with his 'partner-in-crime' Anthony Jones. Since, at the time he started with us in the early eighties the chefs who came to work for us were almost invariably white, two six-foot-two, beefy, black chefs seriously stood out as something different. Mind you, with Ainsley's personality he would have stood out if he had been five foot and grey.

We used to have competitions between the kitchens for 'Kitchen of the Week' and in order to make his team conspicuous Ainsley would deck them out in spectacularly created silver foil hats and scarves. However it was not only the foil that earned Ainsley and Anthony the first prize on many occasions as the food they produced was of an exceptionally high standard.

While I always considered Anthony the better chef, there is no doubt that Ainsley was the motivator. It didn't surprise me when I discovered that after a day's work with us, Ainsley would go off to do a stand-up session at the Comedy Store in Leicester Square with a partner as 'the Calypso Twins'

Also not a surprise was that, when we were away from London at a golf tournament near York and accommodated for the week in a University Hall of Residence, Ainsley and Anthony were the 'Pied Pipers' who led the partying, quite often requiring me the next morning to placate the warden with apologies and prudent amounts of hush-money.

Apologising and dealing with complaints effectively is an art form, using a judicious mixture of humour and humility, that took me years to learn, largely listening to excuses given to me. The most perfect example I ever heard was very early on in my career. We had been catering for a golf tournament in Bournemouth and Judy and I decided to stay on for a few days afterwards at the Dormy House Hotel at Ferndown.

After a while I got very bored with the standard daily menu and asked the young waiter if I could see the 'À la Carte Menu'. His eyes glazed over – quite apparently this was not a common request. I repeated it carefully to him and he went off to see the headwaiter, who obviously asked the lad who had dared to make this demand, as the waiter was pointing to us. It was rather like a Bateman cartoon, 'the Man Who Asked For the À la Carte Menu'. The headwaiter pompously went over to his station, picked up a leather-bound tome, blew the dust off it, and gave it to the lad to bring across to me.

After a few minutes I decided to start with 'Oeuf en Cocotte à la Crème', a simple dish but quite popular in days before puritans, wishing to deny us our pleasures, invented cholesterol.

The waiter said, "Do what?" So I took his finger and pointed it at the item on the menu and he went off with the whole book to the kitchen.

Judy had sensibly ordered Pâté Maison, which arrived quite quickly. I waited, and waited, and waited. Eventually the young man arrived with a very large soup bowl, awash with milk, in which sat two rather desultory looking poached eggs peering up at me.

Horrified, I said to the waiter, "What on earth do you

call that?" To his eternal credit he responded, "That, sir, is the chef's third attempt."

This certainly took the complaining wind out of my sails - and replaced it with laughter.

16

AGA SAGA

During the seventeen years I worked with Gus before he died, one of the regulars at his Elevenses, and usually for lunch afterwards, was a charming man called Buzzer Hadingham. An endearing sprite of a man, possibly half a dozen years younger than Gus. Nobody ever knew, or if they did, never used, his proper name, as he was set for life by his two-year-old brother welcoming the arrival of "my baby buzzer".

During this time he was firstly Managing Director and then Chairman of sports manufacturer Slazenger, and as such a natural visitor to the golf and tennis events where Gus catered. He, understandably, was very keen to maintain the integrity and independence of the Slazenger name despite it being a relatively small part of a large industrial group.

Although Slazenger was never a real customer to Gus and certainly not likely to be as important as its parent, the Dunlop Group, or its sibling rival, Dunlop Sport, sponsors of our largest promoted golf tournament, Buzzer was still treated with the respect his position deserved and lavished with just as much hospitality as Eric Hayes, the boss of Dunlop Golf.

Right from the very beginning while looking after Gus's hideaway I was given very clear instructions that all guests were to be treated the same. This was certainly one of Gus's most endearing traits and certainly against the normal way of business, even in his day, and probably why he was so special to so many. It is one that I have subsequently tried to follow, although not always as even-handedly as he, since, as I pointed out before, not all of Gus's guests were worthy of his bounty.

The impartiality of Gus's largesse was to rebound in a most agreeable way some five years after his death when I received a most delightful letter, which to my chagrin I no longer have, from Buzzer. He stated that for the twenty-five years he had received Gus's hospitality he was never able adequately to return it. He went on to say that since he had just been elected Chairman of The All England Club, he was now able to repay the son in some small measure for the father's generosity. Would Judy and I like to be his guests in the Royal Box at Wimbledon? Would we ever?

So a few months later we found ourselves walking up the steps to the grand main entrance to the Centre Court under the ivy-clad balcony where the All England Club committee members and their guests can look down on lesser mortals. At that time invitees were not, as a rule, given lunch, merely an appropriately English afternoon tea at a suitable break in play.

After a drink on arrival we were shown out into the Royal Box itself and noted from the formal pecking order that we were to be in the third row of Lloyd-Loom armchairs seated between the headmasters of Harrow and Eton – I'm sure an unintended irony by Buzzer. Having refused the proffered blankets (in Wimbledon Purple and Green, of course) to protect our ancient knees from the midsummer climate, we took our seats and looked around the ground to see anyone we knew in the Centre Court – more importantly anyone who could see us, as the great thing about *being* in the Royal Box is being *seen* in the Royal Box by friends who *aren't* in the Royal Box. After the first session of play we had our very pleasant tea and then Judy and I went outside and took advantage of our exalted position to stand on the balcony over the entrance and observe the hoi polloi shuffling past.

Suddenly there came from below a plaintive, "Judeee". It was a near neighbour of ours and fellow member of our baby-sitting group (and so a good way to get the message back). Since they couldn't come in to join us we went down for a chat and it turned out that they had seen us sitting watching the tennis and said, "That can't be Judy and Alan in the Royal Box!" Nice moment – silly, but nice.

A couple of weeks later I had lunch with Buzzer in my tent at the Open Championship at Royal Birkdale and he told me that he had a number of things that he needed to change in the hidebound Wimbledon establishment during his tenure. One of them was to crack the fifty-year plus monopoly of the catering by my father's old employer, J Lyons & Co and its offspring, Town and County Catering. Buzzer wanted to take us in there. "Hear, Hear," said I.

Sadly, while he achieved a great deal in his five years, that was one move too far, and it took financial ructions within Town and County after Buzzer had retired to eventually change things on the catering front, unfortunately not in our favour. Buzzer did however issue a further invitation to the Royal Box four years later, by which time it had been decided to give guests the option of having lunch, which needless to say we accepted.

Thus we once again went through the club portal to be greeted by a uniformed RAF Officer who asked, "Would you like a table to yourselves or would you be happy to join another couple?" I replied that we would be pleased to have some company – networking is networking and this is pretty exalted stuff even if someone else was doing the catering.

We sat down and a few minutes later a very elegant elderly couple joined us, "Lord and Lady Camrose" – "Alan and Judy Payne". Seriously exalted stuff; his lordship and his brother owned the Daily Telegraph. During a delightful lunch I discovered that in the hurricane the previous October, while I lost a whole golf tournament of tents, he lost a thousand trees from his estate, which indicated that his garden was possibly somewhat larger than mine.

I had mentioned the fact that we had just taken on the catering at Goodwood Races, which they said was a course they enjoyed. Then he asked, "Do you do the catering at Ascot?" when I said no Lady Camrose said, "Oh that's good. Don't like Ascot much, although I have to go because my son has horses running there."

Lord Camrose then said, "What's the biggest problem in your business?" I said, "Timing. Just like your newspapers, I suppose. With us it's staff and customers turning up late." He went on, "Tell us about the worst offenders?" I replied that we had recently undertaken a State Banquet for the King of Morocco, who, the first time in the history of the Guildhall had turned up half an hour late. "Wasn't he the wretch who kept the Queen waiting on the tarmac at Rabat?" "The very same." I then went on, "Then there was the dinner we did for the Aga Khan" and before I could say that he had kept us and six hundred Ismailis waiting nearly three hours, Lady Camrose cut in, "Oh, my son."

As conversation stoppers go that was certainly a beaut. Somehow you don't expect the Aga Khan to have something as mundane as a mum, let alone that you would be sitting down to lunch with her. I managed to mumble something slightly less than derogatory about his tardiness and Judy was just about able to finish the mouthful of food that nearly choked her. As soon as was seemly I disappeared to the loo and was convulsed with delayed laughter when Buzzer came in. I told him the story and he joined in the hilarity and agreed that it would certainly have stumped him.

Judy said afterwards that she was glad she had been there, otherwise she would have put it down as just another of my unlikely stories of extraordinary things that happen to me on a reasonably regular basis, all of which she thinks I make up.

The shock of the revelation did serve to remind me of the evening at Guildhall when we helped the Ismaili community in the UK to celebrate the twenty-fifth anniversary of the accession of the Aga Khan, and the build-up to it, all of which caused it to be called within P&G "The Aga Saga" long before Joanna Trollope's novels acquired that nickname. The whole history started at around the same time in 1983 as the Roux Brothers contacted us about hiring equipment for the DaDa evening at the Albert Hall and the preparation for both events carried on in parallel.

We heard from the Keeper's department at Guildhall that representatives of the Ismaili community had hired the Great Hall for a major banquet to be held in July and that they had been given the names of Payne & Gunter and the other four caterers on the Guildhall list for such events.

Sure enough the following day Dr Sachadina, a leading light of the Ismaili community in Britain and the chairman of the team delegated to organise the function, contacted us. I knew relatively little about the Ismailis except for the fact that ethnic cleansing in East Africa in the early seventies had brought a very welcome influx of them to Britain, which had had a similar beneficial effect on business life as the arrival of East European Jewry in the 1890's. Given their success over the previous decade I could only surmise that their departure had an equal and opposite effect in Uganda and Kenya.

From the reputation they had as serious traders, I knew that in spite of our recent run of successes at Guildhall – we had been awarded virtually all the State occasions there for the previous two years – our getting the Aga Khan banquet was going to be very far from a shoo-in.

So we had a meeting with the good doctor and a couple of his colleagues where he laid down their requirements and informed me that, not unexpectedly, he was seeing all the other caterers on the Guildhall list. We certainly rose to the challenge; in fact we relished it. John Coomb and the Head Chef set to work organising a suitable menu incorporating, as requested, both European and Indian cuisine – the latter with help from an Ismaili restaurateur – in readiness for a competitive tasting dinner. I decided to find out what I could about the Ismailis. I was sure that diligent research would pick up some advantage for me over my competitors.

Once again I have to point out the differences between the time I am writing and the time I am writing about. Whilst undertaking this, I have been able to put the word 'Ismaili' into my computer and have available a vast amount of information. Admittedly one has to be able to separate the wheat from the chaff, the bone fide from the downright scurrilous, but it is all there instantly. In 1983 it required a couple of days at various reference libraries and a visit to the newly built, but not as yet officially opened, Ismaili Centre.

While I was able to find out a store of useful information about my potential clients – mainly as a means of not offending their sensibilities – it was the Ismaili Centre itself, a wonderful marble-clad edifice on the island site opposite the Victoria & Albert Museum, that was my inspiration.

It was such a lovely building and was obviously at that moment of enormous significance to their community. Since it must have required a great deal of work and no little fund-raising to make a reality I figured that an ice sculpture of it at the tasting dinner would have the desired effect. So I commissioned Duncan Hamilton, a very talented ice sculptor, to do a very large and detailed model of the building. This cost over a thousand pounds, no little sum in 1983, but I was sure that it would prove a good investment.

It was arranged that Dr Sachadina and five of his committee would come to dinner at British Grove, as it happens the day of my first nightmare meeting with the DaDa team at the Albert Hall, and we had to make sure that everything was at its best.

John Coomb and I were to sit down to dine with them, so our mahogany table was laid up for eight in the grandest style that our Mayfair Catering inheritance could muster. But even all the silk cloths, silverware, crystal and flowers were overshadowed by the second table we had introduced just inside the entrance. This had on it the magnificent ice carving, approximately four-foot square and three-foot high, of the beautiful Ismaili Centre. Outstanding enough by itself but it was made totally breathtaking by having spotlights pointing down on it as well as lamps within the ice itself that made the whole thing glow an iridescent blue.

Having satisfied ourselves all was in readiness, we wanted to postpone the 'coup de theâtre' until all the preliminaries had been dealt with. Since it was a fine, warm spring evening we greeted them all in the car park where I had welcomed Simon Roberts a few months earlier. In order to delay the denouement, we took them on a tour of the warehouse and kitchens and it was there that I sought to establish the form of the evening.

"I know that alcohol is not permitted by your religion, but I am not sure how rigidly this is observed. In order not to cause offence, could I ask for clarification? We have wine available as well as a variety of fruit juices and waters. Will any of you have any wine? Would you have any objection to John and me having wine while dining with you?"

Dr Sachadina said, "You are absolutely right, alcohol is forbidden by our beliefs". Chuckling slightly he went on, "You are also right in implying that observance is not always absolute, possibly even by some of us. However, since this evening is an official occasion for our community, none of us will have any alcohol but we have no problems with you taking wine alongside us. We very much appreciate your courtesy in asking us." We had been walking through the warehouse while talking and had reached the door to the Pantry, John opened it and ushered them in ahead of us.

The effect when they walked in and saw the floodlit ice carving, with the rest of the room lit only by two candelabras on the dining table, was one that will live in my memory as one of the great moments in a career studded with highlights. Initially it was the joy that we had obviously done the right thing professionally to achieve our goal. Several minutes went by before any of them said a word as they walked round and round the carving, marvelling at it with open-mouthed admiration. The realisation of the exquisite pleasure that I had given them by identifying their pride and joy, and honouring it in this way, made my evening as well as theirs.

The meal that followed, while exceptional and very well received, was inevitably somewhat of an anti-climax, particularly since they found it difficult to take their eyes off the carving. One of the bright spots for me during dinner was when Dr Sachadina let drop the fact that the chairman of one of our competitors, who had dined them earlier in the week, had greeted them with the words, "Would you like a sherry?" When told that the Ismailis do not drink alcohol, he responded, "Well I think we have a Bitter Lemon." Without my guest saying who the competitor was I was aware, due to the industry grapevine, who had hosted them previously and I have to say that a company with almost as long a history as Gunters should have known better.

The subsequent negotiations centred not on whether we should undertake the dinner but at what price. They were somewhat hampered by the fact that they knew they wanted us to do it, and we were buoyed by the fact that we knew they wanted us to do it. I had done my negotiating, setting out my stall in the Pantry.

"Mr Payne, do you realise that your price is ten pounds a head more than your nearest competitor?"

"I didn't."

"Are you surprised?"

"Not really."

"Are you not going to say anything to help me?"

"You have an opportunity to save your community six thousand pounds. I am afraid that you have to make up your own mind whether they will thank you afterwards for doing so."

He grinned at this, acknowledging the irony for a community of traders. The discussion was very good humoured as he appreciated that he was fighting a losing battle. I knew from the moment they entered the Pantry that the dinner was ours – my only regret was that by that time we had already quoted our price, otherwise he would have really had something to complain about.

I had no qualms about holding out for a good price – not that I ever did – because, as well as knowing that John Coomb would do a better job than anyone else, if the standards they attained in the building of the Ismaili Centre was anything to go on, these were going to be very particular and demanding clients. Sure enough their attention to detail, while extremely satisfying professionally, created an enormous amount of work.

The community had decided that the occasion was so special that each guest, as part of their ticket price, would be able to take home the complete cover of china, silver and glass their dinner was served on. These were specially designed and manufactured for them and delivered to British Grove for us to take to Guildhall. Each course was designed for minimal food to be left on the plate - for example, no bones - so that after each course the guests could put the china and cutlery into a specially provided bag to take home. The only person not to take the equipment home was, of course, the Aga Khan himself as his cover was whisked away and was auctioned later.

We had weeks of detailed trials of dishes both at British Grove itself and at Indian restaurants with our head chef trying to perfect the Tandoori Chicken, which was to be the second course, after a fine dish of salmon that we had specially line-caught and smoked for us in Scotland. The salmon was to be very simply served with a half lemon wrapped in muslin so that the juice could be squeezed out without pips covering the fish. This was to be followed by a tea sorbet and then Beef Wellington, with fillets of beef, which our Edinburgh butcher would have hung for ten days prior to shipping it down to us for the occasion.

Since there was obviously not to be any wine, we had a variety of different fruit juices and waters. As we had no idea which of these would be popular and which left alone we had to have very large quantities of each; given the subsequent delay this was no bad thing. I was also inclined to overdo the ordering of the juices after a near disaster at the very first State Banquet we did at Guildhall a couple of years earlier.

That dinner had been in honour of King Khalid of Saudi Arabia and while we had, as instructed, made provision for orange juice for all the Muslim guests we had not made allowance for all the other guests to eschew wine and go for the soft option. As a consequence we ran out of orange juice within minutes of the dinner starting and had to send vans out to get orange juice from wherever they could, which required some initiative in the City of London in the evening.

Some while later I was relieved to see the cellar master, who normally would have been busying himself with decanting fine clarets, frantically opening large cartons of fruit juice. My relief was short-lived as it turned to abject horror when I noticed that each carton was emblazoned "Produce of Israel".

The dinner we were producing for the Aga Khan at Guildhall was the Grand Finale of a day of activity including an exhibition at Olympia and others at the Ismaili Centre. Timing was going to be critical as after his arrival at Guildhall there was to be some ceremony in the Old Library before the banquet itself.

All the preliminaries, trials and tests had been undertaken during a period of intense catering activity including creating a brand-new golf tournament at a course we had never been to before in the centre of Glasgow, the very last Martini Golf tournament after twenty-five years, and the British Masters under a new sponsor, our old client Dunlop Sports having pulled out of their fortieth event only weeks before. As well as this there was the pantomime with the Roux Brothers at the Albert Hall to contend with.

The fact that I had insisted on the chefs taking a disproportionate amount of time over the Ismaili dinner while all this was going on had given it a very high significance to the kitchen brigade.

On the evening itself all the guests arrived in some confusion because many of them had been in attendance at some of the other functions around London. This was compounded by the most monstrous rainstorm, which flooded parts of West London including Olympia where the Aga Khan was. Again, the whole thing would have been more easily managed had we had mobile phones but we, and four or five hundred Ismaili guests who had turned up, were completely in the dark about what was happening as six o'clock passed to seven and then to eight with more guests trickling in but no guest of honour.

For John and me the main activity, besides wandering out to look after the, by now, very fractious guests, was placating the chefs who were beside themselves at the thought of all their preparation resulting in a devastated meal. Luckily enough, when the Aga Khan did arrive they decided to carry on with the pre-dinner ceremony, which gave the kitchen time to cook the Beef Wellington to perfection.

The only notable happenings during the much-delayed dinner itself was trying to placate our frazzled chefs when all the guests decided to undo and remove the muslin coverings on the lemons that had taken so many fiddly hours to tie up before the evening. They also were less than delighted by the fact that the Ismailis had so taken to the ways of their adopted country that they demanded HP Sauce with the Beef Wellington, a request we couldn't immediately satisfy. As a summary of the evening I couldn't improve on what our head chef said at the time, while performing miracles as a cowboy in the kitchen – moving from range to range – "It's bad enough cooking *for* an Aga, imagine if we'd been cooking *on* one as well!"

17

1786 AND ALL THAT

When we took over the fine old business of Gunters in 1980, it occurred to me that we could have some fun with an invoice we discovered, though I wasn't quite sure how. It was dated 1786 and made out to The Lord Peterborough, from 'Negri & Gunter at the Pot & Pineapple in Berkeley Square' for the princely sum of one pound sixteen shillings.

In the immediate aftermath of the takeover of Mayfair, having to contend with State Banquets, arriving at RFU Twickenham, near bankruptcy and our own takeover by Jobs, following up on the history of Gunters quite reasonably took a back seat. However by 1983, with our new partners installed and additional management brought in, I thought it was time to start looking into this, as I didn't want to miss the opportunity that the anniversary in 1986 would bring.

We had already taken aboard the date of the invoice as a tongue-in-cheek piece of chutzpah for the foundation of the company. We had a small scroll as part of our letterhead stating that we were 'Established in 1786' (although, of course, Negri & Gunter must have been in business far earlier), it also appeared in the 'Everything Goes Down Rather Well with Payne & Gunter' adverts in the Sunday Times.

In the latter part of 1983 I set a couple of researchers to work, one on the history and one on the menus of Gunters to see what, if anything, we could make of it commercially. The information they found out, added to my informal discoveries of the previous couple of years, was very exciting. The Gunters had been an amazing family.

We discovered that one member of the family had taken as his inheritance some of the company's market garden and developed it, and so created a large part of what we know as Chelsea today. Gunter Grove, Tregunter Road, Redcliffe Square and Gardens and The Boltons were all part of the Gunter Estate. That one of the Gunters went on to be an MP and a baronet was quite something in the snobbish Victorian era for someone whose family had been 'in trade'.

It also emerged that the famous confectioners had held the Royal Warrants from George IV through to George V, which doesn't sound a great deal except, of course, that it also includes William IV, Queen Victoria and Edward VII - quite a record. The fact that they also catered for Louis Napoleon and Eugenie when they were living in England at Farnborough Hill, and subsequently when he had become Emperor Napoleon III, was evident from the Warrant that was handed over by Mayfair and which now hangs in my loo.

Progressively we found out about great social and state events for which Gunters had catered from the late eighteenth to early twentieth century together with the menus and gradually we put together a very detailed dossier of our findings. The question now was, what to do with it?

The few years after the takeover of Mayfair had seen a considerable increase in the banqueting activity of the company, initially in the livery halls and Guildhall in the City, under the leadership of John Coomb and Bernard Clements and subsequently in some more interesting and unusual locations.

An extremely talented catering manager called Roger de Pilkyngton, who subsequently went on to become our sales and marketing director, had driven this process. Roger was able to convince the trustees and managers of some of the country's most prestigious buildings that letting parts of the properties out for private parties for fees of many hundreds or thousands of pounds for the odd evening was going to cause considerably less wear and tear on their precious charges than thousands of feet of the great unwashed.

As with most of the changes which we initiated, this has now become so much the norm and such a large part of the income of most museums, palaces and stately homes, that it is hard to remember just how difficult it was only twenty odd years ago to persuade a very sceptical group of people that we were not Visigoths or Viking raiders about to pillage their treasures.

Such wonderful buildings as Queen's House at Greenwich, Hampton Court Palace and the British Museum succumbed to the blandishments of Roger and the cheques of his customers. They were also reassured by the respect shown to their buildings by all the Payne & Gunter staff, almost without exception (indeed any exceptions were very severely dealt with – usually by flogging or hanging, sometimes followed by drawing and quartering - to be fully in keeping with the history of some of the buildings).

We also used some more eccentric venues, including the Toy Museum, where Roger organised a party for Simon

Robert's fiftieth birthday, the scene being completed by model trains running round the table carrying 'Smarties' in the wagons with staff dressed in school uniforms and Roger as Circus Ringmaster with top hat and riding whip.

That the guardians of these buildings were less than worldly-wise is illustrated by the very first function we undertook in Apsley House. This is the wonderful Regency building on Hyde Park Corner that his grateful nation built for the Duke of Wellington after Waterloo. Rejoicing in the address, No 1 London, Apsley House is still used in part by the current Duke but the house and its magnificent art collection are open to the public.

Although it had been designed and built to stage grand banquets, it hadn't been used for that purpose for many years. That was until Roger persuaded the V&A, who managed the house, that he was to be trusted. The insurance company who wanted to celebrate with a magnificent banquet were very suitable first tenants as they even pre-dated the Iron Duke, and were prepared to hand over a very large cheque for the honour.

A reconnaissance visit showed Roger the room we were to utilise as a kitchen and servery that, because it contained some wonderful pictures, could not have anything in it that produced steam. The electric turbo ovens for the beef would be all right, but not the soup or new potatoes.

After checking the electricity supply for the ovens, (he did this very carefully as a few weeks before, at our very first function at the Natural History Museum, our staff plugged the electric ovens in, tripped the electric fuses, and killed all the butterflies), he went downstairs to see where we could do the rest of the cooking.

It was decided that since there was a space to park next to the current Duke's space – I imagined the first

Duke tethering his horse, Copenhagen, there – we would set up Calor gas rings in the back of the van for the soup and vegetables.

On the highly stressed day of the banquet, Roger arrived to discover that our parking space had been usurped by an enormous pile of gravel, ready for some maintenance work. An Apsley House functionary said, "Don't worry, the Duke is at Stratfield Saye today, so it's safe to park in his place."

Everything was set up and going smoothly, with the guests having arrived and enjoying a tour of the house while drinking their Roederer Crystal Champagne, when a frantic flying functionary flung open the servery door and proclaimed, "The Duke's coming, move the van!" Protestations had no effect, "If you want to do any functions here again, move the van."

So, with three lads in the back holding on to the soup and potatoes, the van began a very gentle tour of Hyde Park Corner at seven o'clock of a Friday evening. A taxi-whistle from Roger and the van came in for the soup to be served and off it went again, until it was time for the tatties. The hosts and their pampered guests remained blithely unaware of the unorthodox peregrinations of their soup and potatoes. However, the near calamity meant that we ensured that they found us alternative facilities to properly undertake the many subsequent banquets we did there.

To encourage customers to have functions in these fantastic locations, Roger started to get us involved with the worldwide incentive business. This market, primarily in America, but becoming popular all over the world, relies upon companies wanting to offer prizes to their star employees as a way of motivating them. In order to supply these, a large number of agencies vie for the companies' business, offering a vast variety of prizes for the successful staff members.

These range from simple gifts to days at the races, theatre nights, weekends away, full Caribbean Holidays and for the very top star performers, tailor-made tours for groups of winners to interesting and prestigious locations. The more exceptional or unique the venue and the greater the exclusivity of it (in other words, the greater the impossibility of the hoi polloi getting in) the more the incentive winner will value the prize.

We got to know some of these agencies very well and a couple of years after the Apsley House function we had a dinner for a number of them including the one whose client had been hosting that dinner. In my speech afterwards I related several hair-raising stories, near disasters all, about our events including this one. The agent concerned went very pale as it dawned on him that it was his dinner, "I didn't know all that was going on while we were having that dinner." Quietly I said to him, "Did you want to?"

During the 1980's we developed more and more venues and offered them to agencies who would arrange a complete week in Britain for, as an example, the hundred top sales achievers for a life assurance company based in the US. This would include five-star accommodation and travel, top seats at impossible-to-get-into shows in the West End and lunches or dinners in palaces or other locations denied to the general public. It was extremely successful and a fascinating new market for us.

It was not lost on Roger and me that this market was where the Gunter heritage was going to come into its own. We gathered together all the information regarding dishes that we knew Gunter had served to each of the monarchs for whom they held the warrant and with help from some old-established wine merchants such as Deinhards, Veuve Clicquot, and my old friend the wine merchant from Chichester, Chris Purchase, we were able to serve a complete menu of dishes and wines which Gunter had served to, for example, George IV. We also had appropriate menus for the other crowned heads including the Emperor Napoleon.

Having established the right food and wine, I then used the history and social background research to create a script for each of the menus so that each course could be introduced with information about when it was served and anecdotes about the Gunter family and business.

A wonderful friend, Colin Turner, who was at that time deputy head master of Latymer Upper School, helped me in this task. I had first met Colin over thirty years previously, when he was a student teacher and I - as an eight-year-old new arrival at the prep school at Latymer - was one of his first pupils. Novices together. He it was who instilled my love of words, and subsequently mentored my children Emma and Richard, particularly encouraging them to participate in the drama activities, which he organised, and for which the school had become famous.

Some twenty years after leaving Latymer, when I purchased the old warehouse in British Grove in Chiswick for Paynes Caterers to move into, I wrote to all the residents of what was effectively the mews behind the grand houses of St Peter's Square, Hammersmith, to reassure them that the warehouse site was not going to be redeveloped as they had feared. (When, subsequently, our neighbours experienced the comings and goings of our vans, lorries, trailers and staff mini-buses at every odd hour of the twenty-four, with the warehouse burglar-alarm going off what seemed like every other night, they probably regretted the fact that the development didn't happen.)

Colin Turner lived in one of the houses in British Grove, and so we renewed our acquaintance. His home became a regular watering hole in the evenings after some of the frenetic days I had during our time at Chiswick, where he introduced me to the joys of Gonzalez Elegante Fino Sherry – rather a lot of Gonzalez Elegante Fino Sherry.

It was with the help of Colin, and doubtless the sherry, that the scripts were completed and we decided to try one of them out at a particularly important occasion for Payne & Gunter. We had just won the contract for our first managed hall in the City. The Worshipful Company of Tallow Chandlers had been looked after by Ring & Brymer for many years at their Livery Hall in Dowgate Hill, but feeling rather neglected, the Tallow Chandlers Company had put the catering out to tender and our bid succeeded. We wanted to put on a very special dinner to celebrate, not just for their senior liverymen but also for other actual and potential users of the hall.

We therefore agreed that, since there was a stained glass window looking out onto a courtyard with a portrait of Edward VII, we would use the menu of food and wine served by Gunters to King Edward. We also decided that we would employ John Coomb as 'Pageant for the Evening' reading the script that Colin and I had prepared for the occasion.

John, while a very self-confident character, was not used to speaking in public and avoided doing so. It took quite a bit of persuasion before he agreed. He was very nervous when we rehearsed him in his script and given the nature of the man, he made a joke of it including taking some liberties with the lines. At one point during the section when he was to introduce the Salmis of Pheasant, we had given him the words, "King Edward was, of course, a renowned shot" before going on to describe the gargantuan nature of Edwardian meals. The first time John read it out to us in the Banqueting Hall of the Tallow Chandlers' Company with its Grindling Gibbons carvings around the wall, he, inevitably, said: "King Edward was, of course, a renowned shit."

This brought the predictable gales of laughter from all the members of staff who were trying to finish laying up the tables. Try as we might, each subsequent reading produced the same effect as if John had, inadvertently, got hooked in a loop. I just hoped that he would get it right later on that evening.

The evening was a great success, starting with the guests' arrival past the discreet wrought-iron gateway from Dowgate Hill along a passage under some offices, which opened out into the charming colonnaded city courtyard with a single catalpa tree providing a dappled canopy. After entering through the impressive double doors to the Hall under a stone carving of the City of London shield, they were in the foyer, panelled in highly-polished dark wood with the coats-of-arms of all the Masters of the Livery Company for the last five hundred years emblazoned round the cornice. In the well of the grand staircase, we had installed a half-size replica of the magnificent cake, with ornate icing depicting angels standing guard over all Her Imperial Majesty's Colonies and Dominions, that Gunters had produced for Queen Victoria's Diamond Jubilee as a gift for Her Majesty. The original had stood ten foot six high and weighed a quarter of a ton.

The dinner went absolutely according to plan and John's act as pageant was well received and thoroughly enjoyed by all as a novel way of getting diners to actually think about what they are eating. Not surprisingly I was on tenterhooks as the pheasant course approached, the dishes for the previous course having been cleared from the table and with John striding out into the centre of the hall and loudly proclaiming: "Ladies and Gentlemen! King Edward was, of course a renowned ...... shot". The rotten bugger had had me on the go the whole time.

So with the success of the first of the Gunter Bicentenary Banquets at the Tallow Chandlers Hall behind us we now decided to publicise the concept in a really big way. In the summer of 1986, the chosen bicentenary year, we were catering for the second year of a successful new Pro-Celebrity Golf Tournament called the Four Stars. Using a similar format to the popular, but questionably accounted, Bob Hope Tournament, the Four Stars who fronted the event were Henry Cooper, Bruce Forsyth, Jimmy Tarbuck and Terry Wogan and this time the organisers did hand over some considerable funds to the charities involved.

On its first year the Four Stars had a Gala Dinner the night before the tournament at the Grosvenor House Hotel, which Payne & Gunter organised and administered but of course did not cater for. For the second year I suggested that we should organise and cater for the dinner and stage it at Guildhall and that we should serve a Victorian Banquet to celebrate our actual Bicentenary.

Thus it was that I found myself seated on the top table along with the four celebrities and their wives and a few seats along from the Princess Royal whose charity was benefiting from the evening. The major drawback of sitting on a long top table rather than on round tables, as the other guests in front of us were, is that if the people seated on either side are talking to the guest on their other sides, you are left with no one to talk to and nothing to do. Unless, of course, you are a caterer.

What I did when those moments came, was to watch all the round tables being served, flinching at every tiny mistake being made by our waiting staff, totally unable to remedy all these transgressions. Inevitably a fault caught my eye – a waitress serving wine from the wrong side right across the customer's plate - so blatant that I couldn't resist reacting. I scribbled a note on my place card and instructed the waitress serving us to give it to John Coomb.

A few minutes later, the suave John Francis Coomb shimmied Jeeves-like behind me and whispered in my ear, "I had noticed, Sir. Here are a few more place cards just in case anything else attracts your attention." I was put in my place beautifully. And I had to stay in my place and suffer manfully.

Some months before this dinner I had managed to get a first edition of a wonderful book entitled 'Gunter's Confectioners Oracle', dated 1831, written by William Gunter, the doyen of caterers during the first few decades of the nineteenth century. He had remained running the business while his brother James had joined the Dragoons and subsequently built Chelsea. One of the great things about the book, besides it being a marvellously eccentric guide to the joys of eating, including a chapter on The Philosophy of the Stomach, was that on the frontispiece was a drawing of the very louche figure of William Gunter. From then on, in place of a 'Pageant' describing the dishes and Gunter anecdotes, we would have William Gunter himself. An actor was hired for this role at the Four Stars Victorian dinner and played the part with aplomb and the whole production was greatly enjoyed by a potentially very critical audience, even by me in my perch on the top table.

Now that we had a proven formula to sell, it was time to package it and present it to the market. Roger de Pilkyngton had pulled together a dozen of the prime venues we used for banquets such as the ones we wanted to sell. These properties, including the Tallow Chandlers Hall, Hampton Court Palace, Natural History Museum, Brigade of Guards Museum, The Queen's House at Greenwich and Syon House, now that they had overcome their initial resistance to the idea of the commercial use of their facilities, were happy to pay to have us represent them around the world and bring in suitable groups of visitors.

Since one of the great joys about these buildings was the exclusive and secret nature of them, we decided to market them on the basis that we were able to let the client in on a very private world to which only we possessed the key. The joint venture was called, appropriately, 'Great British Secrets', and given the role it was to play, suitably stylish printed and display material was produced.

The company, represented by Roger, then appeared at various exhibitions and conferences all around the world dedicated to the incentive market. Poor Roger had to flog round this dispiriting list of places such as Geneva, Hawaii, Cancun and Costa Rica – staying always at the five-star hotels that hosted these events; he was so noble, he never complained, not once.

The only times that I managed to get in on the act was for the main get-together for the industry, the Incentive Travel & Meetings Exhibition, which takes place each year in the exceedingly unromantic environs of McCormick Place Exhibition Halls in the warehouse district of Chicago. It was agreed that I would appear in the role of William Gunter, dressed accordingly in order to maximise the publicity for our mission.

So it was that Roger, soberly dressed in a simple grey suit, and his managing director, dolled up to look like a

Regency Dandy, drove each morning of the show through the Chicago rush-hour and set up our very large stall as an important part of the British Pavilion. We were an important part because, besides exhibiting, we had agreed to organise food and drink of a suitable nature for the entire British contingent. On arrival at McCormick Place we set about our separate tasks, Roger to detail our catering staff and then to sell our wonderful venues and banquets to visiting representatives of the incentive agencies from around the world. Most of them of course he already knew as fellow travellers who had shared with him the deprivations of all those mundane places.

My job was to be hawked round the massive exhibition by Barbara Boyle, head of the British Tourist Authority team, as a living, breathing example of historic Britain. Besides this rather enjoyable role of 'stage prop' I also had the opportunity to be interviewed by various local and international television crews, for which fortunately I had had some coaching and rehearsing before I left London.

I got to know one of the TV producers quite well and met him for a drink at a bar across the road from the exhibition hall. Apparently it was considered sensible to take a cab the short distance to the bar, as the area was a bit run-down to say the least and more than a little dangerous for visitors. As I didn't know this, I strolled across the road and wandered into the very pleasant bar. On my entry dressed in tailed frock coat, frilled shirt, cravat, drainpipe trousers with spats, top hat and silver-topped cane, the old black pianist with grizzled white hair who had obviously seen it all, from Al Capone onwards, stopped playing, turned round and growled at me, "You come over here dressed like that! And you're still alive?"

The producer was from a Chicago Station and, after

a few beers, he told me a tale of a previous station he worked at in Washington DC. One December they spoke to their contact in the British Embassy to ask what he would like for Christmas. Being very British he protested, "I'm sorry, I couldn't possibly accept anything." The TV station presenter insisted, "Aw go on - you've been very good to us over the past year." Cautiously the diplomat agreed, "Very well, perhaps a small box of crystallised fruits."

When Christmas Eve duly arrived, the envoy was watching the local television station when the announcer said, "We did a sounding of the Embassies to see what the diplomats want for Christmas and the man from the French Embassy said that he wanted greater cultural understanding between our two countries. The German representative wanted more economic cooperation between the US and Germany and the Brit wanted a box of crystallised fruits." When I, laughingly, said I didn't believe him he said, "It's as likely to be true as you are to get back across to the exhibition in one piece." Since I did, I suppose I have to believe him.

The producer also told me that he had interviewed the pianist after he discovered that he was over ninety and still working all hours. When asked what the secret was for his long, healthy life he responded, "Had I known I was going to last this long, I'd've taken better care of myself."

One evening in Chicago, Roger and I took Barbara Boyle from the BTA out for dinner to a place recommended by the TV producer. It turned out that the restaurant had been booked for a large Golden Wedding celebration of a local Jewish couple, Issy and Rachel. However, mention of the producer's name allowed us to have a table at the side of the dining room.

Not surprisingly we got talking to guests of the party

on adjacent tables and inevitably the question arose, "So how come you English guys are here at Issy and Rachel's do?" Wickedly, and with a poker face, Roger said, "Barbara, here, has been Issy's mistress in London for the past twenty-five years." The effect on the party as this story went round was magical but not surprisingly we had to leave quite soon, trying not to bust a gut.

After we got back to London with our arms full of orders, we had to fulfil the commissions we had earned and one of them produced more than a little difficulty and the inspired creation of an 'instant tradition'. We had been involved in many of the ancient ceremonies of the City of London, such as the Swan Feast of the Worshipful Company of Dyers, who, as one of the two City Companies entrusted for many centuries to look after the Queen's swans, are allowed to serve one at their annual ceremony.

We also undertook the catering for the Worshipful Company of Cutlers and their eight-hundred year-old ceremony of the Boar's Head Feast at which a company servant called the 'Beadle', in grand medieval robes, leads the procession of a choir followed by two chefs holding shoulder height a wooden stretcher, carrying the decorated Boar's Head, followed by a third chef holding carving knife and fork aloft. Having offered the dish to the Master of the Company, the Beadle in his flowing vestments is given an orange on his salver, which he is bid by The Master Cutler to give to the leader of the choir.

The current Beadle told me that the first time he did it, as he turned to the choir, the orange rolled off the salver and disappeared under the table. In subsequent years, the Master has carefully placed the orange on a bit of Blu-tack. I've no idea how they coped for the last eight hundred years without modern technology.

It now appeared that through a set of circumstances to do with twentieth century traffic we ourselves were going to add to the City's traditions. We had sold the idea of a Georgian Banquet to a Dutch bank and I was going to play the part of William Gunter, describing the wonderful occasions when I had served the dishes to His Majesty King George IV at his various palaces including Brighton Pavilion. The dinner was going to take place in Gibson Hall in the City of London, which is quite as over-the-top as Brighton Pavilion. Initially called Natwest Hall it was a Victorian Banking Hall that was the City Branch of one of the companies that made up Natwest. Surplus to their needs, they re-decorated it in garish late Victorian style and started to hire it out for banquets. They changed the name to that of the architect because, quite reasonably, all the other banks, which made up a large part of their potential users, didn't like dining chez Natwest. We had used the hall quite often and it featured in our list of Great British Secrets.

So, dressed once again as a Regency Dandy, I got in my car in time to drive from Richmond to the City with Roger already there setting up the dinner, which he was going to oversee. As soon as I started, I stopped. West London was in total traffic gridlock, but although this was in the early days of cell phones, I used mine to inform the office that I would be delayed, and for them to pass this on to Roger. They told me that Roger had everything in hand at the hall but no candles for the candelabras. A car was dispatched from British Grove with these while I tried going further and further round South London to find a way to get to the City as I didn't fancy going on a rush-hour tube or bus in my Regency garb.

The car with the candles got stuck in the same traffic

jam and was eventually abandoned in favour of the tube. Meanwhile I was going yet further round London, through Brixton and getting more than a little worried about what would happen to me if I had to abandon my car – if there is somewhere less kindly than the warehouse district of Chicago towards a Regency Dandy, it's probably certain parts of Brixton.

There was Roger, sans William Gunter, who never got there at all, sans candles, but ready to serve this elaborate banquet. In a dining hall that gave one the impression of dining inside a giant multi-coloured wedding cake, he had a couple of hundred expectant and highly valued guests of the host bank, flown in for the occasion, mightily impressed with their surroundings, and anticipating the works.

Since Roger knew the scripts as well as me, he did a fine job of performing the pageant, but without costume, while managing the catering. While the fish course was being enjoyed, the candles arrived and were put into the silver candelabras.

With inspired inventiveness, after the clearance of the first course, Roger called for silence with his gavel and grandly announced, "My Lords, Ladies and Gentlemen, will you please be upstanding for the Ancient Ceremony of the Arrival of the Candelabras." At which point everyone stood up and rhythmically clapped as the waitresses walked in, each proudly carrying a candelabra with lighted candles and placed one on each table; the Dutch merchant bankers were duly impressed with the old City of London customs. Wonderful thing, tradition.

18

Eight Hundred Years On

It was a dull November day in the bleak backrooms of Guildhall and it was our turn to go into the committee room to face the grilling - caterers are always grilled, in the same way, I supposed, as bricklayers are laid-off and cobblers are always last. Although I had been through this process a few times, I was very much the novice compared to my colleague, John Coomb, who led the way into the room, and who was without doubt the seasoned old trooper.

We stood before the splendour of a City of London Reception Committee – a dozen of the Great-and-the-Good of the City (their description). The function for which we, along with three other caterers, had received an invitation to tender was 'the big one', 'the plum job', 'the goody'. It was the one we all desperately wanted.

Put simply, the Corporation of the City of London were holding a lunch to celebrate the Eight Hundredth Anniversary of the Mayoralty of the City and, since these do not come around too often, they had invited the Queen to be their guest of honour. 'Put simply' was most certainly not the way it was going to be.

Not only was it a 'to-die-for' invitation for the guests but it also caused a less-than-decorous scramble amongst the dignitaries of the City to get onto the reception committee. More pertinently for me, it would be the way to establish our company, Payne & Gunter, as absolutely the prime caterer in the City of London a mere ten years after John joined me and we undertook our first City banquet. Quite an achievement - should we get it. That was why John and I were facing this ordeal, he with considerably more equanimity than me.

My nervousness was, to a great extent, assuaged by the presence of John beside me; he had after all experienced more of these meetings than any of the committee, this being 1988 and John having worked at his first Guildhall banquet, albeit in a much more junior role, in 1948. He started his well-practised introduction and looked the embodiment of calmness and comforting security.

He stood beside me, both of us in black jacket, waistcoat and striped trousers, he four inches taller than me, slim, rigidly erect, with stiff white collar and silver grey tie, the same colour as his very full head of hair, in contrast to my bald pate and beard. There was always a chuckle trying to creep out of the corner of his mouth and a sparkle in his blue eyes that had any ladies on the committee eating out of his hands. As one of our Liverpudlian drivers put it so succinctly, "He always looks the dog's bollocks".

The committee might not have put it quite that way but John certainly gave them the one thing they so desperately required - reassurance. He would deliver and had delivered many more times than anyone else.

Three weeks prior to this we had submitted written presentations, a very elaborate and grand dossier, with a copy for each member of the committee detailing everything they could want to know to help them come to the sensible conclusion that Payne & Gunter should be given the honour to undertake the catering for the lunch.

We listed the staffing numbers and organisation, how they would be deployed, and where we would source them; the front-of-house management, with photographs and histories of each significant player, and of course a detailed review, with pictures, of the kitchen brigade led by our head chef.

Having been given a brief of what could and could not be served to the Queen we put together four menus, including appropriate wines, at four very different prices, each commensurate with the vastly differing costs of ingredients, most especially the wines.

We were then, along with our competitors, summoned to appear at Guildhall to be interviewed by the committee. Guildhall is not just the magnificent Gothic hall that appears on television at the time of the Lord Mayor's Banquet, when the prime minister addresses the City, it is also the massive bureaucratic heart of the City of London, for centuries the mercantile capital of the world.

We were all instructed to be in attendance in the anteroom by two o'clock. This always gives the merry band of caterers an opportunity to gossip. We knew for a fact that the committee would not have finished their lunch for at least forty-five minutes, and then would shuffle their papers for a further quarter of an hour, so it was going to be three o'clock at the earliest before the representatives of the first company would be ushered in.

It did relieve the tension to chat with one's competitors, even if we all knew that each of us would do whatever was required to win. Since most of us were old friends, past colleagues and sparring partners of many years' standing, there was a pleasant atmosphere of banter as we set about discussing the capriciousness of the Corporation's decision-making for each banquet.

Then came the moment when the first one was called into committee and he naturally became the topic of conversation for the remainder – I have often been tempted to leave a recorder in the waiting room and have frequently wondered whether any of my competitors ever have.

So eventually the Grand Vizier of the Corporation, who rejoices in the title of The Remembrancer of the City of London, called John and me into the committee room. He was a very imposing chap, six feet five or so, and except for his suavity, he could easily have been a second-row rugby-forward. He certainly was mightily impressive, given the enormous arrogance of his office, even without his full-bottomed wig and gown, which he would duly wear on the occasion of the lunch itself.

After several minutes' questioning and some discreet discussions within the committee, the Remembrancer announced in his powerful baritone, "The Committee feels it will go with your menu B." This was not a surprise because Menu B was by far the cheapest of those on display. Having agreed Menu B, and again to no one's surprise, least of all ours, a member of the committee made a suggestion, "Would it be possible to substitute the Délice of Sea Trout from Menu D for the pâté in Menu B?"

This is the beginning of a City Tradition that has been going probably as long as the mayoralty itself. It is known in the square mile as 'Screwing the Caterer'. All members of the committee, whether friends of the caterer or not, will subsequently propose similarly helpful substitutions of food and wine until not one jot of Menu B remains save, of course, the price.

Once this was successfully agreed we returned to the waiting room, where our rival fellow caterers had been waiting patiently (quite appropriate since most of them started their business lives as waiters) for their turn to suffer the Grand Inquisition.

After the last candidate had returned to join his peers, there was the interminable wait while the committee deliberated. The next part of the procedure would be for the Remembrancer to emerge and peremptorily dismiss the unsuccessful candidates. He would then invite the chosen one back to establish the date for what might be the first in a succession of tastings at the caterer's premises where a few, or indeed all, of the Committee would try the various dishes. This would also involve, in the case of several professional committeemen, as many wines as possible.

However we at Payne & Gunter had anticipated such an event and had over the years gathered in some very special parcels of wine adequate to serve 700 or so guests. We had unleashed some of these on the waverers; Chablis Grand Cru Vaudésir 1984, Château Gruaud-Larose St Julien 1978, Veuve Clicquot White Label Rich, Cockburn 1975.

I believe that the 1975 Cockburn was the clincher, as a true Vintage Port had not been served at a Guildhall Banquet for the best part of twenty years; it was obviously a blandishment too far. The committee allowed itself to be corralled by the Chairman into agreeing that Payne & Gunter should be given the honour.

Having established that we would be undertaking this task, after allowing ourselves a glass or two of bubbly to celebrate, there followed several weeks of intense activity planning the logistics and ensuring that everything and everyone arrived at Guildhall as and when intended. Given the long drawn out preamble, one can realise the pressure on the day itself for John Coomb to perform. It might have been his hundredth state occasion for the City, but the same logistical problems that applied to all of them applied equally to this one, even more so, because the expectation was higher. The tension of the occasion, which the anticipated presence of the Monarch always ratchets up a notch or two, and the demands of our client, the Corporation, communicated itself to even the most humble member of John's team and I was not the most humble member of John's team; in fact I was, of course, his boss.

Although I had been with him at the committee meeting and the tastings, it was John's reputation and reassuring presence they had chosen, but should anything go wrong it was my company's reputation that was on the line. I had to let him run it, and no one better, but at each of our special events all of the top team at Payne & Gunter took relatively junior positions, supporting whoever happened to be in charge that day; I was no different; rather like an Admiral visiting a ship and deferring to the Captain.

I therefore helped with the staff allocation and, at the function itself, took responsibility for the top table. For such an occasion it was vital that we had the right service staff. This was not an occasion for a waiter with a three-day-growth and Tourette's Syndrome or waitresses who have taken as instruction the first Queen Elizabeth's comment that she "took a bath every month whether needed or not" (I must admit, to my chagrin, that both have slipped through the net at some time or other).

However, this was the time for the best.

Altogether we had to have some hundred and twenty waiting staff and twenty chefs, together with thirty other staff behind the scenes, and a management team of twenty.

Every single one was known to us and had been cleared by security checks.

We had arrangements with various catering colleges over the years and each summer I created a competition to decide the best college for the year that, as a prize, would be allowed to bring a dozen of their top students to work at one of the very special occasions for which we would be catering during the forthcoming winter.

So it was that on this particular day in February a minibus made its way from Rotherham to London carrying the twelve best students from the local catering college, which had been considered to be Payne & Gunter's best educational partner the previous summer, to play their part in this alien pageant in the City of London. Ten of the students would join the main team on the Champagne reception and subsequently serve wine in the great hall. The two star students would be delegated to me and be part of the elite team looking after Her Majesty and the rest of the top table.

I had visited Rotherham College the previous December and talked to the students about the occasion, the history, Guildhall and all other aspects of the event. It would appear later that I didn't include all aspects; one omission would bring us close to calamity.

After a tension-mounting security check at the gate, the minibus was allowed into the underground car park at Guildhall so the nervous, excited and very apprehensive team of northern lads and lasses came through the labyrinths under the historic building. Through the kitchens in the basement, where twenty chefs were hard at work at various stages of preparation, up the service stairs to the enormous servery where at this time, some two hours before the off, it was already a scene of tumultuous activity. Plates were being stacked in hot cupboards; literally hundreds of service spoons and forks were being laid out interwoven so that, during the peak of service, waiting staff could grab a pair on their way out to silver-serve the guests in the hall.

At the end of the servery the students went through the mighty room called the wine bin where dozens of bottles of claret and port were being prepared for decanting. Half the decanters, those for the port, were being blessed with a discreet elastic band round the neck, to avoid confusion with the claret, and the Chablis and Clicquot were being chilled and made ready for service.

Because they had entered Guildhall through the 'below-stairs', the students' nervousness and excitement visibly increased on seeing the frantic activity, without them seeing the object of it. This was not to be until after they had changed in the staff rooms into their smartest whites, starched and pressed by their proud mums the day before.

When they were ready I took the students and their lecturer on a quick tour of Guildhall. Even twenty-eight years after my first visit, and ten years of being regularly involved in these functions, I still found Guildhall, prepared for a State Occasion, a jaw-dropping, awe-inspiring place, and I have always enjoyed showing people the spectacle for the first time and watching their reaction. I was certainly not disappointed this time.

The granite corridor leading from Guildhall Yard to the Old Library was carpeted for the occasion and completely bedecked with flowers on both sides, up to the ceiling and three feet deep. Banks of every imaginable flower and plant grown by the Corporation's own Parks Department. (The first time I heard this, I had a vivid image of all the City's parks being denuded for the occasion until I discovered that the Corporation have hundreds of acres under glass in Lea Valley.)

As this corridor is over 200 feet long, the effect was staggering, especially in a London February, and the aroma totally overwhelming and bemusing. It crossed my mind that a visiting Head of State suffering from hay fever, would never make it to dinner.

We then moved in to the Old Library, also festooned with flowers and basking in the warm strains of a baroque quintet. Readied for the Champagne reception, it made a dramatic contrast compared with its mundane weekly use as the Court of Common Council, the citizens' local parliament for the last thousand years.

The City has gathered to itself a vast array of gold, silver and crystal over the centuries, as each Lord Mayor, from Dick Whittington onwards, typically presented something to the City at the end of his term of office. These, together with gifts from visiting heads of state during eight hundred-plus years, means that when the Corporation wants to pull the stops out they are really able to do so.

Guildhall itself was laid up for 600 guests. Nearly a millennium's worth of the City of London's finest gold platters, rose bowls, candelabras, urns, vases, loving cups and goblets was displayed throughout the hall. This, together with our best silver, crystal and gold and white china, laid out on the finest white napery, freshly ironed on site, made a seriously impressive array.

Red and white roses (in February), a foot high, adorned the entire length of all the tables and were also in great cascades behind the top table and from the galleries. Palm trees were in each corner and tropical plants in every nook and cranny, with massive ancient silk banners from all of the City's hundred and ten Livery Companies hanging below the Gothic roof. Union Jacks and the white and red flags of the City of London, as well as those of England, Scotland,

Wales and Northern Ireland were high up on the walls. The whole thing made for an astounding sight - and one not often seen in Rotherham.

I gave the wide-eyed students a brief résumé of the occasion and timetable of events and then asked the lecturer to identify the two chosen to assist me while the others went back into the servery to be allocated tasks. My two young Yorkshire stars came with me, as it was time to look after my area of concern, the top table, where they were to be the runners.

Typically at State Banquets - if any such occasions can be called typical - the top table runs the full length of the central section of the long south wall turning into sprigs at each end and thus forming a very large horseshoe, seating perhaps seventy or eighty people.

When the monarch is guest of honour, the top table is on the raised dais at the short eastern end of Guildhall and seats no more than 20. Since this is the only table on the dais, it reduces the total number for such an event to 600 from the normal 730 for a State Visit or Lord Mayor's Banquet.

The other major difference is the method of service as traditionally the Queen is not silver-served from the platter by the waiting staff, as are other guests. She is offered the dish so that she can decide how much, and to what, she helps herself, which is what I had first experienced all those years ago at Moët's chateau. It is referred to as 'butler-service' as it was the standard form of service in Victorian and Edwardian houses, certainly in the heyday of our Gunter predecessors.

The twenty people in my care were to be very well attended.

To look after their security, two City of London policemen in their dress uniforms, looking to all appearances like cast-offs from a Gilbert & Sullivan opera, stood, still as statues, behind the top table – nothing moving but their eyes.

For the wine service we had brought in a couple of the most experienced butlers from two of the oldest of the City's Livery Companies, the Worshipful Company of Vintners and the Worshipful Company of Skinners.

The food was to be served by Payne & Gunter's two finest waitresses, who on a daily basis looked after our own 'Pantry' and who therefore had served at the tasting lunches that won the contract. They were to be assisted by my two students who, as runners, did precisely that, to and from the servery. They also had to put down the plates for each course and serve the vegetables after the waitresses had served the meat.

I then gave the final briefing to my team for the Top Table, which was now my absolute priority. From this point on I had to forget about the rest of the hall, which John and our other directors were managing.

My troops were each allocated their positions and I then went through the menu detailing how each course would be dished-up and how one waitress and her student would start with Her Majesty and go south and the other with the Lord Mayor and go north. I also pointed out that the Queen, Duke of Edinburgh, Lord Mayor and Lady Mayoress would be butler-served and everyone else, silver-served.

Now it was time for the reception to take place in the Old Library where the remaining ten students, among others, were serving Champagne. For my team it was a chance to sit down quietly, or perhaps not so quietly, in the staff rooms, awaiting the call to arms.

For me this is always the most wondrous moment; Guildhall itself is empty and silent, the normal echo is stilled by the drapes and unaccustomed carpeting; the tranquillity is sublime, although I know only too well, it is very much the calm before the storm.

We knew exactly how long it took to dish-up each course in the kitchen and transport it in the heated lifts up to the servery.

We were kept informed intimately of the toastmaster's timetable. Past experience told us that from the moment he calls the guests in to dine, it would take exactly fourteen minutes for 600 people to progress down the ambulatory and find their seats in Guildhall (quite amazing, but even the most difficult people in the land do precisely what they are told when they are told when dining with the Queen).

At appropriate times during this process our staff were gathered from various sitting rooms around Guildhall and held in the servery, 'armed' and ready for the fray.

I heard the State Trumpeters' fanfare and the announcement of the royal arrival, and it was now that I had my quiet moment of panic, as I knew that in precisely seven minutes the first course must be served.

Then, at the 'moment juste', with a precision that would have delighted Busby Berkeley, the elegant and unflappable John Coomb sent in his team.

We're off. My top table team moved into action.

The students brought in the heated fishplates and laid them before the guests and the two waitresses followed with the Délice of Sea Trout Belle Meunière; the City butlers poured the Chablis Vaudésir and the comic-opera policemen stood vigil.

The fish course eaten, the plates, gratifyingly empty, were cleared and taken away by the students, who returned with the hot plates for the meat and placed them before the guests. The Saddle of Veal with Apricots was served. The students took the Gardeners' Pride round the table and the crystal decanters of Château Gruaud-Larose were proffered by the white-tied butlers. I was starting to relax.

For the first time I allowed my eyes to stray to the rest of the hall and I felt a quiet sense of satisfaction and elation at the efficient bustle of our staff serving 600 of some of the most important people in the land. I thought that we had indeed 'cracked it'.

The staff had almost finished the ends of the table when I happened to walk past one of the policemen. From under his walrus moustache, out of the corner of his mouth he said something I have never forgotten and still wake up in a cold sweat re-living, "Queen ain't got no veg."

I looked and he was quite right, the Queen, the Duke of Edinburgh, the Lord Mayor and the Lady Mayoress all 'ain't got no veg'. No time to question why – filled with horror and foreboding, I was in the servery and back with some Gardeners' Pride in seconds - offered it, with as much dignity as the required speed allowed, to the neglected honoured guests and hosts and I think only Her Majesty noticed the hiatus and, of course, she was far too well brought-up to mention it.

When I asked the students after the lunch was all over why it was that they failed to serve the Queen with vegetables, the response, in flat Yorkshire vowels, was: "You told us that the Queen and the others would be 'butler-served' – why didn't the butlers serve them?" - Simple really. However, it isn't true that there is never a policeman around when you need one.

19

LORD MAYOR'S SHOW – AND AFTER

A year after nearly depriving Her Majesty of her greens and it was my turn to host a grand banquet at Guildhall. Given my rudimentary early days when a visitor to one of our golf tournaments could have anything they liked for lunch as long as it was steak and kidney pie (and in those days it was a considerable novelty to be served anything hot) it was heady stuff when, at the end of the seventies, I started catering at grand banquets in the magnificent historic halls of the City of London. The idea of even being a guest at one was unthinkable never mind hosting one and being applauded by six hundred people onto the top table at Guildhall to a fanfare by the State Trumpeters.

I gradually became aware of my surroundings, I heard the toastmaster's gavel right behind me, "Alderman, Ladies and Gentlemen, Pray silence for Mister William Gunter, by appointment confectioner to Her Imperial Majesty, Queen Victoria."

At which point a portly gentleman in frock coat,

sporting mutton chop whiskers, stood up at the next table and proceeded in a deep baritone to tell the assembled guests of the occasions he had served the various dishes and wines, which made up the banquet they were about to enjoy, to Queen Victoria at her various palaces.

This entire enterprise was, of course, a confection of my own making. Stemming from the takeover of the Gunter name a decade before and the discovery of the rich history it involved, I was now using it for a very special and appropriate purpose. It was very special, because the evening was to raise money to help restore the delightful Georgian folly built by James Wyatt on Temple Island in the Thames between Henley and Marlow at about the time that Mary Gunter set up shop with Domenico Negri in Berkeley Square. It was appropriate, because the banquet was to celebrate the 150th anniversary of Henley Royal Regatta, which had just been given ownership of the island, which forms the start of their historic course.

As managing director of Payne & Gunter, who were not just the caterers but also impresarios of the whole evening, I was going to be the host, along with Peter Coni QC, the chairman of the regatta.

My involvement with Henley had started fifteen years before, when we first created a hospitality pavilion operation at the Regatta to help fill the coffers that Richard Goddard, the incoming Secretary, had discovered were virtually empty when he took over the role. Our venture had been a resounding success, to the considerable benefit of the Regatta and ourselves, and this evening was, in part, to celebrate a further five-year contract.

I took the time while 'Mr William Gunter' was introducing the first course, as he was to do with all the subsequent ones, to look around the magnificent building and the sixty-five round tables. We had sold over five hundred places at the dinner to our corporate customers and we gave fifty places to Henley for their own guests. We had taken the opportunity to show off what was certain to be an exceptional display of food, wine, service and its history to invite a hundred of our own guests from all areas of the company's activity, as well as friends and family. As I absorbed the scene, I realised that in this massive room this evening, the assembled throng represented virtually every aspect of my life, both business and personal; the only one missing was Gus.

When 'Mr William Gunter' sat down at the table on my immediate right, after his initial introduction, next to him was Emma, my nineteen-year-old daughter, absolutely gorgeous in a buttercup yellow dress that would not have been out of place at a ball at the end of the very first regatta in 1839. Alongside her was Richard, her brother, younger by two years, looking uncannily like me at the same age, sneaking an evening off from his 'A'-level exams. Then there was their grandmother, my mother Lila, still very stylish at nearly eighty. It was appropriate that they should be on 'William Gunter's' table because the actor playing that role was not an actor at all, he was in fact head of English at Latymer Upper School, where I had my education and Richard was just finishing his.

Sitting on 'Mr William Gunter's' right, was the man who had played a very important role in the lives of both of my children and me. Colin Turner, six foot four inches tall, with a full head of silver-grey hair, very elegant and suave with matinee-idol good looks, was deputy head master at Latymer, and the driving force for twenty years of their fine reputation for drama. This evening he was doing what was he was best at, eliciting a good performance out of the 'Victorian Confectioner' sitting next to him, having spent the previous two months helping me to fine-tune and rehearse this evening's script, inevitably over a great many glasses of his favourite Gonzalez Elegante Fino sherry.

Colin had thoroughly enjoyed the nineteen-eighties, revelling vicariously in the successes of two of his pupils, Mel Smith on television and Alan Rickman in the theatre. Tragically, within a few months of this dinner, Colin was to die, accidentally falling down stairs, ironically the first Saturday lunchtime he had not visited the pub for many years. He was not to see the subsequent Hollywood blossoming of Rickman or the meteoric career of his third star pupil, Hugh Grant.

While the soup was being served, I glanced round my table and, as I looked at one of my guests, I realised that although Colin had given me a love of the use of language he had obviously left some gaps in my knowledge because, for the life of me, I couldn't think what in English was the opposite for the phrase 'a shrinking violet'. Whatever it was linguistically, there on my left, chatting to my wife Judy, was a fine example in the flesh – the very ample flesh - of Sir Peter Yarranton, six foot five and 19 stone, bald as – well, bald as me – our guest of honour for the evening and, as a member of the Rugby Football Union Committee, the first to introduce me to the delights of Twickenham.

The fact that Peter was in all senses 'a character' was well known to anyone in the world of sport, whether observing him bellowing like a bull-moose from the roof of the grandstand of Wasps Rugby Football Club at a hapless winger who had committed some minor infringement, or listening to one of his many fine after-dinner speeches (fortunately this particular evening he was going to be following me) or hearing him recount, as a priceless excuse for arriving late for a dinner, the fact that he had had to stand in for Lord Denning, the senior Law Lord, at a reception – even his excuses were larger than life.

I only realised how much he was in a class of his own with regard to self-promotion when, prior to this dinner, he had given me a copy of his CV so that he could be introduced correctly, and it ran to five close-typed pages of career, achievements, charities, clubs, honours and associations. He stood out even within the RFU, an organisation bursting with large and noisy men.

I continued gazing round the table while the soup plates were cleared. To Peter's left was another old rogue (progress through this book must have shown that I have an affinity for old rogues, although, to be fair, most were young rogues when I first met them) by the name of John Clevely, a wine merchant of some note, old fashioned, tall and distinguished, who could give the impression on first meeting of being a pompous old fart, however nothing could be further from the reality. While he has passed the exams for the highest accolade of the wine trade, he insists on describing himself as "Jack of all trades and Master of Wine".

For the previous 20 years or so he had been responsible for importing Veuve Clicquot Champagne into the UK and when, sixteen years before, I had been looking for help from various suppliers to fund my trial run of a hospitality pavilion at Henley Regatta, he was the only one of the Champagne houses to offer any. This, justifiably, earned him, and Clicquot, a very profitable monopoly for more than a couple of decades.

The most memorable thing about him is that in the 40 years I have known him – we first met when I was working for my father at the tennis tournament in Bournemouth during my school holidays and John was a young van-boy delivering beers and wine – he has almost always been late, to the extent that he is forever known in the wine trade as "the late John Clevely".

The trouble with such a reputation is that everyone gets to depend on it; so I arranged a surprise 50th party for him on the day after his birthday – it just seemed more appropriate. So much did we rely on his tardiness that one morning he was due to come down from his office in Wapping to see me in Chiswick at eleven o'clock and I carried on working at my desk on the assumption, born of long experience, that he would not be with me until one o'clock at the earliest.

This was in the days before mobile phones and I was, therefore, more than a little astonished when he turned up at eleven with a worried expression and the perfect rejoinder to the inevitable, "What are you doing here already?" that greeted him. "Well I was trundling westwards on the District Line, realising that I was likely to arrive at eleven and I didn't quite know what to do. I was going to stop and phone my secretary to get her to warn you. The trouble was that I wasn't sure about the etiquette of her calling you to say, 'I do apologise but Mr Clevely is going to be on time'". Delightful man.

'Mr William Gunter' was back on his feet introducing the fish course and I looked across my table to Richard Goddard, one of my fellow hosts, and saw that he was thoroughly enjoying the script I had written; I hoped that he would approve later of the fact that I was to use his appearance and most especially his walrus moustache in the speech I was to deliver at the end of the meal.

I had no real worries on that score, as Richard is someone who enjoys everything, not least laughing at himself. I specifically remember, around the time of the opening of the new regatta headquarters, Richard bursting in with a copy of the local newspaper, spluttering with mirth. He had been interviewed by a young journalist and had made the comment that, "the Regatta is a curious animal", which is no more than fact, given its unusual format, its longevity and, most importantly, the social context of the event. The front page of the paper rejoiced in a very large picture of the Secretary sitting outside the new regatta offices with the heading, "Richard Goddard – a curious animal".

'Mr William Gunter' was just doing his last introduction and the audience did not want to let him stop. I realised that it was nearly time for my turn. I have always considered myself a competent speaker, rather than an inspired one, and although I have often had to get to my feet and address an audience I have never lost the extreme nervousness the anticipation creates. However, this was of a different magnitude altogether; six hundred and fifty people in a massive Gothic hall which had resonated to the voices of Lord Mayors and visiting Kings, Queens and Presidents for hundreds of years, and the Prime Minister of Great Britain every November at the Lord Mayor's Banquet.

I had to tell the audience of the generosity of one of the Regatta Stewards, Alan Burrough, who, having recently sold his family business, Beefeater Gin, had purchased Temple Island and given it to the Regatta on a peppercorn lease – at a rent of two eggs per year – not my rotten scotch eggs I hope.

Following that I had to welcome everyone and propose the toast to the Regatta on the occasion of its 150th year. Since I knew that I was going to be followed by two of the slickest, amusing and most accomplished speakers around, I decided to do a bit of doggerel. On account of the appearance of the Secretary of the Regatta and the calling of the Chairman of the Regatta and with due apologies to the estate of Lewis Carroll, I offered a pastiche that I titled 'The Walrus and The Barrister'

While standing up talking to six hundred people on the spot where the Prime Minister lays out his foreign policy to the nation every year, I couldn't help thinking about what Gus would have made of it all. For all his effervescence and bravado, Gus had been a very reluctant formal speaker, even to a small audience of friends. There was one time when we made a droll feature of this inability at one of his annual lunches for his golf sponsors. After the meal had been served, I rapped the table for attention and said, "Gentlemen, pray silence, the chairman, Gus Payne, would like to get to his feet."

Astonishment rippled round the table as the guests, most of whom by this time had known him for many years, had never, ever, heard him make a speech. Gus stood up, looked around for several seconds, took a couple of thoughtful puffs on the Ramon Alones and slowly sat down again - to tumultuous applause.

I then stood up, said, "Thank you very much, sir" and carried on with my speech as usual to the guests.

Without doubt there was a very large element of 'After the Lord Mayor's Show' about the next few years of my life following that celebration of Henley Regatta's 150th birthday. After all, the term has come to mean 'it's all downhill from here' and to an extent, the Henley Banquet at Guildhall was a high-water mark in my life.

However, the Lord Mayor only comes into his own estate after the pageantry of the show itself, when he steps out from the shadow of his predecessor. I also felt, after that dinner, that I had nothing more to prove to Gus. I had, in any case, developed the business dramatically since his death and taken it into areas he wouldn't have dreamt of, most of them highly profitable.

So, in the summer of 1989, after hosting my grand Henley dinner, I really was Alan Mark I rather than Gus Mark II – and not even Little Alan any more. I could now look upon Gus as a very lovely memory, rather than a looming presence.

Post Script

John Coomb passed away in September 2006 - this book had nearly been completed but sadly he had not had a chance to read it. His funeral was, as might be expected, a jolly affair, rejoicing in the fun of the man while recognising the enormous loss we all felt. It ended up in the President's Room at RFU Twickenham, which, along, with Guildhall in the City of London, was one of John's spiritual homes for more than fifty years.

Gathering my thoughts to give a short address I realised that there could be no better coda for this book than recognising the two beliefs that John and I shared and which made our partnership such a merry one; that you owed it to your customers not to mess up *their* special occasion and, more than anything else, that serving people should always be fun, for those doing the serving as well as those being served.